Algebra: For High Schools and Colleges

Ellen Hayes

Copyright © BiblioLife, LLC

BiblioLife Reproduction Series: Our goal at BiblioLife is to help readers, educators and researchers by bringing back in print hard-to-find original publications at a reasonable price and, at the same time, preserve the legacy of literary history. The following book represents an authentic reproduction of the text as printed by the original publisher and may contain prior copyright references. While we have attempted to accurately maintain the integrity of the original work(s), from time to time there are problems with the original book scan that may result in minor errors in the reproduction, including imperfections such as missing and blurred pages, poor pictures, markings and other reproduction issues beyond our control. Because this work is culturally important, we have made it available as a part of our commitment to protecting, preserving and promoting the world's literature.

All of our books are in the "public domain" and some are derived from Open Source projects dedicated to digitizing historic literature. We believe that when we undertake the difficult task of re-creating them as attractive, readable and affordable books, we further the mutual goal of sharing these works with a larger audience. A portion of BiblioLife profits go back to Open Source projects in the form of a donation to the groups that do this important work around the world. If you would like to make a donation to these worthy Open Source projects, or would just like to get more information about these important initiatives, please visit www.bibliolife.com/opensource.

ALGEBRA

FOR HIGH SCHOOLS AND COLLEGES

BY

ELLEN HAYES

PROFESSOR OF MATHEMATICS IN WELLESLEY COLLEGE

Norwood Press
J. S. CUSHING & COMPANY
1897

PREFACE.

PART II. of this work is a revision of the author's *Lessons on Higher Algebra* prepared several years ago in view of the need of a suitable text-book for a brief course in algebra required in the freshman year at Wellesley College.

In developing a short course in higher algebra, it has seemed desirable and has been found possible to give unity to the work by proposing one general problem, — the determination of the roots of higher numerical equations. Most of the matter of Part II. will be found to bear on this problem.

It is the author's experience that students of the mathematics usually offered in college need to have an elementary algebra at hand for reference. Part I. has been prepared to meet this need. It is also believed that for the purposes of high schools and preparatory schools, Part I. will prove a helpful supplement to the ordinary algebra text-books, since it emphasizes certain important aspects of the subject which at present are being more or less neglected. In writing this part, two classes of secondary school students have been kept in mind: those whose mathematical studies will end with the high school course, and who are entitled on that account to know something of the spirit and methods of the mathematical fields which they are not going to enter; and secondly, those who expect to advance to mathematical work of a more difficult grade, and who ought not to find an abrupt break between the elementary course of the preparatory school and the higher work of the college or university.

ELLEN HAYES.

WELLESLEY, MASS , Jan. 1, 1897.

CONTENTS

PART I

CHAPTER I

ARTICLES		PAGE
1–18.	ALGEBRA AS A LANGUAGE	1

CHAPTER II

| 19–23. | ADDITION AND SUBTRACTION . . . . | 15 |

CHAPTER III

| 24–33. | MULTIPLICATION . . . . | 20 |

CHAPTER IV

| 34–38. | DIVISION . . . . | 28 |

CHAPTER V

| 39–46. | THEORY OF EXPONENTS . . . . . | 35 |

CHAPTER VI

| 47–58. | SURDS . . . . | 44 |

CHAPTER VII

| 59–64. | IMAGINARY QUANTITIES . . . . . | 52 |

CONTENTS

CHAPTER VIII

ARTICLES
65–70. FACTORS 56

CHAPTER IX

71–88. EQUATIONS 65

CHAPTER X

89–94. RATIO 81

CHAPTER XI

95–105. PROGRESSIONS 88

CHAPTER XII

106–114. INEQUALITIES 93

CHAPTER XIII

115–117. VARIATION 101

CHAPTER XIV

118. REVIEW 105

PART II

CHAPTER XV

119–131. DERIVATIVES 111

CHAPTER XVI

132–139. RATIONAL INTEGRAL FUNCTIONS 123

CHAPTER XVII

ARTICLES   PAGE
140–141. BINOMIAL THEOREM 135

CHAPTER XVIII

142–145. CONVERGENCY OF SERIES 140

CHAPTER XIX

146–153. THEORY OF EQUATIONS 146

CHAPTER XX

154–156. GRAPHS 157

CHAPTER XXI

157–168. SPECIAL ROOTS 164

CHAPTER XXII

169–175. STURM'S THEOREM 175

CHAPTER XXIII

176–186. DETERMINANTS 189

CHAPTER XXIV

187–204. THEORY OF LOGARITHMS 200

CHAPTER XXV

205–210. MATHEMATICAL REASONING 215

PART I.

CHAPTER I.

ALGEBRA AS A LANGUAGE.

1. When we consider two or more things and attend merely to the fact that they are distinct one from another, we may *count* the things by applying in order the words one, two, three, etc.

For instance, if we have a handful of roses some of which are large and red, others small and white, and others medium sized, very fragrant and pink, we disregard differences of size, color, fragrance, etc., and simply register the repeated impressions made on our senses by these objects as distinct objects; having completed the registration we declare the *number* of the roses.

If now we attend to one of the roses and describe it as a large rose, we are considering it quantitatively and are roughly *measuring* it by means of a rose of medium size which is our unit of measure. Perhaps we choose a medium-sized pink rose from the handful and with it as a unit of measure say that the red rose is twice as large as the pink one. In doing this we have applied the unit to the *quantity* to be measured, and counted the number of times that this application could be made without applying the unit more than once to any one portion of the quantity.

To illustrate further, if a bin contains an unknown quantity of wheat, and we find upon trial that a bushel basket can be filled fifty times in succession before the wheat is all removed from the bin, we declare that the bin had fifty bushels of wheat in it. By counting the times the basket was filled we have been able to count and state as a number the distinct bushels of wheat composing the quantity of wheat. If at the fiftieth filling of the basket, we find the basket only half full, we say that the basket has been filled forty-nine and one-half times; what we obviously mean is, that the bin contained forty-nine and one-half bushels of wheat, and that we found this out by counting the forty-nine times that the basket was filled, and then calling the fiftieth time a half time because the basket was only half full.

2. From the above examples and similar ones it appears that number relates to the question: *How many?* and quantity to the question: *How much?* also, that the question, *how much*, is answered by first noting *how many* times a measuring unit is contained in the quantity to be measured; also, that '*more*,' '*less*' are necessarily suggested both by number and quantity.

It is to be further observed that we may not know anything about the thing to be measured except that it is quantitative, that is, measurable. For example, we may come across the expression "980 dynes"; in the absence of all other knowledge in the case we at least know that a quantitative something which can be expressed in terms of dynes, whatever a dyne is, has been found to contain one dyne 980 times; that is, the quantity is composed of 980 dynes.

3. Arithmetic deals with the relations of number, and hence also with the relations of measured quantity, when the numbers or quantities are denoted by figures.

Elementary **algebra** is arithmetic generalized and extended, — the generalization being secured by the use of letters as symbols of quantity. When any letter as a or n or x appears in an algebraic expression we simply read the name of the letter; but we are to keep it in mind that when we say 'a' or 'n' or 'x,' we mean: 'the quantity (or number) represented by the symbol a,' 'the quantity (or number) represented by the symbol n,' etc.

This use of letters is purely conventional, "a star or the picture of a dragon-fly would serve just as well if as easily made and read."

Besides symbols of quantity, algebra employs symbols of operation, of relation, and of aggregation. (Arts. 5, 7, 8.)

4. Euclid in his *Elements* (-300) stated certain truths or principles (κοιναὶ ἔννοιαι) which long before his time had been found to govern the relations of quantity; of these principles the leading ones are: "Things equal to the same thing are equal to each other," and "The sums of equals are equal." Certain other truths or laws* (Arts. 10 ⋯ 14), taken in connection with Euclid's *axiomata*, form the basis of algebra.

The terms 'more,' 'less' imply two fundamental operations which may be performed on quantity: Increase and diminution. All secondary operations will be found to be combinations of these primary ones.

* For a discussion of the term "law" as used in science, see Pearson's *Grammar of Science*, Chap. III. The term must be used even more guardedly in mathematics than in science.

4   ALGEBRA.

We have, then, quantity symbolically represented, operations symbolically indicated, and processes carried on in conformity to a few laws. Algebra is thus a language devised to express thought, and if we use the term in its larger sense as including the calculus it will be found that this language enables us to follow trains of reasoning with a power and precision not afforded by any other language device. The translation into an ordinary language as English, French, etc., of the formulas developed in algebraic processes affords propositions which are often theorems of great importance, and the theorems will in many cases contain implicitly rules for other operations.

5. Symbols of operation. The principal symbols of operation are: $+$, $-$, $\times$, $\div$, $\sim$.

The symbol $+$ (*plus*) denotes addition; it belongs to the quantity before which it stands, and shows that the quantity is to be added to whatever precedes it. If there is no preceding quantity, it then merely shows that the quantity to which it belongs is a *positive* quantity. In the absence of any sign the plus sign is to be understood.

The symbol $-$ (*minus*) denotes an operation the opposite of addition; the quantity before which it is placed is to be subtracted from whatever precedes it. If there is no preceding quantity it denotes that the quantity to which it belongs is *negative*.

The symbol $\times$ (*multiplied by*, or *times*) denotes that the quantity preceding it is to be multiplied by the one following it. Multiplication is also indicated by a dot ($\cdot$). Thus $a \times b$ is the same as $a \cdot b$. A still more common way of indicating multiplication is to write the symbols of the factors in juxtaposition without any connecting sign. If the quantities a and b are to be multiplied together we write: ab.

The symbol $\div$ (*divided by*) denotes that the quantity preceding it is to be divided by the one which follows it. A more common way of indicating division is to replace the upper dot by the dividend and the lower dot by the divisor. Thus $a \div b$ is the same as $\frac{a}{b}$, and means that a is to be divided by b.

The symbol $\sim$ is used to denote arithmetical difference as distinguished from algebraic difference. $a \sim b$ means that whichever is the smaller quantity is to be subtracted from the other.

6. Positive and negative quantities. The classification of quantities as positive and negative is a feature of algebra. Illustrations best show what is meant by the terms. If we begin the counting of the hours of the day at twelve o'clock noon, three hours past noon would be $+3$ hours, and two hours before noon would therefore be -2 hours. If we begin the counting of the years with the birth of Christ, any year A.D. is positive and any year B.C. is negative. For instance, Euclid lived -300. If distance eastward be regarded as positive, distance westward is negative. If a person walks m miles due east, and then turns and walks n miles due west, he is $m + (-n)$ miles one side or the other of his starting-point; whether he is on the east side or west side of that point depends on the relative arithmetical magnitudes of m and n; that is, if m and n are viewed as signless quantities, and m is greater than n, the man is on the east side of his starting-point, but if m is less than n, he is on the west side of it.

7. Symbols of relation. The principal symbols used to show quantitative relation are: $=, \equiv, >, <, \propto$.

The first of these is read: '*equals*,' and denotes that the quantity preceding it contains the chosen unit of measure exactly as many times as the quantity following it contains that unit.

The triple bar, $\equiv$, denotes identity and may be read: '*is the same as.*'

The symbols $>, <$, denote inequality. Thus $a > b$ means that the quantity a is greater than b. The greater quantity is written at the open end of the symbol.

The symbol $\propto$ denotes variation in magnitude. Thus $a \propto b$ is read: '*a varies as b.*'

The symbols $=, \equiv, >, <, \propto$, are evidently verb symbols; whenever one of them occurs in an algebraic expression the expression itself is some proposition written in algebraic shorthand.

8. Symbols of aggregation. The symbols of aggregation are: the parentheses ( ), the brackets [ ], the brace { }, the vinculum ‾‾, and the bar |.

Any algebraic expression written within the parentheses, brackets, or brace, is to be taken as a whole; and any operation indicated by some symbol outside of the symbol of aggregation is an operation to be performed on the whole aggregated quantity. For example, $a - (b + c - d)$ signifies that the whole quantity $b + c - d$ is to be subtracted from a. A horizontal bar, ‾‾, over an expression serves the same purpose as the parentheses, etc. Also, a vertical bar is sometimes a convenience for indicating aggregation. Thus the forms:

$a - (b + c - d), a - \overline{b + c - d}, a - \begin{vmatrix} b \\ c \\ -d \end{vmatrix}$, all have the same signification.

ALGEBRA AS A LANGUAGE.

9. In measuring a bin of wheat, the number of bushels is found to be the same, in whatever manner we take the wheat out; in measuring a piece of land or of cloth, it is immaterial at which end we begin; in counting a handful of roses, we may start with a white one or a red one, and proceed in any order. Experience has shown that in counting a set of objects *the number is independent of the order of counting.* This may be regarded as the **first law of number.**

10. Suppose now that we have a handful of roses and other roses in a vase; and suppose that by counting the two sets separately we find that there are seven in the handful and ten in the vase. If we wish to know the number in the two sets combined, there are two ways of proceeding: we may first take the seven roses and then add the ten roses one by one; or, we may take the ten roses in the vase and add to them one by one the seven roses in the hand. But by the first law, it follows that the number we arrive at as belonging to the combined groups of roses must be the same, whichever of the two processes we use; that is, seven increased by ten is just the same as ten increased by seven. Arithmetic enables us to express the fact briefly in this way, $7+10=10+7$; but the shorthand of algebra not only expresses the fact for this particular case but for all like cases. That is, if a represents any number and b represents any other number, the law regarding the sum of a and b, when stated in the language of algebra is $a + b \equiv b + a$.

This is known as the **law of commutation** in addition; it amounts to saying that *the sum is independent of the order of adding.*

8 *ALGEBRA.*

11. If b itself is the same as $c+d$, we may replace b by $c+d$, so that the statement, $a+b \equiv b+a$, becomes
$$a+(c+d) \equiv (c+d)+a \equiv c+d+a;$$
but by the law of commutation,
$$c+d+a \equiv a+c+d \equiv d+a+c;$$
also, $\quad d+a+c \equiv (d+a)+c;$
hence, $\quad a+(c+d) \equiv (d+a)+c.$

The principle governing these identities is **the law of association** in addition; it expresses the fact that *the sum is independent of the grouping of the quantities to be added.*

12. If we have a vases and each vase contains b roses, we may find the whole number of roses in two ways: we may count the roses in one vase and then add, by counting, one by one, the roses in each successive vase. We then have as the whole number of roses
$$b+b+b+\cdots \text{ to } a \text{ terms,}$$
and this we briefly indicate by ab, meaning that b has been multiplied by a; or, we may take one rose from the first vase, one from the second, and so on to the last, when we shall have a roses; and because there are b roses in each vase, this operation must be repeated b times; and now the whole number of roses is
$$a+a+a+\cdots \text{ to } b \text{ terms,}$$
that is, ba. Whence it follows that $ab \equiv ba$. Further, let us suppose that there are a vases in each one of e rooms. It is obvious that we shall have the same number of roses, whether we take all of the roses in one room and add to them, by counting, all of those

ALGEBRA AS A LANGUAGE.

in a second room, and so on, as indicated in symbolic writing, $ab + ab + \cdots$ to e terms, that is, $e(ab)$; or whether we take one rose from each vase in each room, and go repeatedly the round of the rooms until all the roses are counted and indicated by

$$(a + a + \cdots \text{ to } e \text{ terms}) + (a + a + \cdots \text{ to } e \text{ terms}) + \cdots$$

to b of these terms; that is, $b(ea)$. We have then

$$e(ab) \equiv b(ea);$$

whence it appears that the parentheses make no difference in the result, although they do indicate a difference in the process by which the result is attained. It also appears that the order in which the quantities are multiplied together makes no difference in the result. Thus we have what is known as the **law of association** in multiplication: *a product is independent of the grouping of factors;* and the **law of commutation** in multiplication: *a product is independent of the order of multiplying.*

13. Finally, let us suppose that the set of b roses in any one of the vases is made up of c white roses and d red ones, so that we have $b = c + d$. The total number of roses is unchanged on this supposition, since qualitative distinctions are disregarded in counting. Then from the statement, $ab = ba$, we have as equally true,

$$a(c + d) = (c + d)a.$$

But since b is made up of the groups c and d, we may, if we choose, allow color groups to determine our mode of group counting. Thus, we may take a white rose from each vase until all the white roses are counted. We then have as the number of white roses

$$a + a + a + \cdots \text{ to } c \text{ terms,}$$

that is, ca; and similarly for the red roses we have
$$a + a + a + \cdots \text{ to } d \text{ terms,}$$
that is, da. The total number of roses is the sum total of the white roses and the red ones; that is, $ca + da$. But the number was previously found to be $a(c + d)$; hence, $\qquad a(c + d) \equiv ca + da.$

This is known as the **law of distribution** in multiplication.

14. Suppose that a quantity consists of a number of equal factors, say m factors, so that it is expressed by
$$a \cdot a \cdot a \cdots \text{ to } m \text{ factors};$$
and suppose this quantity is to be multiplied by another quantity consisting of n equal factors, each factor being a. Then we have as the final result,
$$(a \cdot a \cdots \text{ to } m \text{ factors})(a \cdot a \cdots \text{ to } n \text{ factors}).$$
But by the law of association, this product is independent of the grouping of the factors, and therefore we may obtain the true result by multiplying $\overline{m + n}$ factors together, introducing the factor a at each operation. We have then
$$(a \cdot a \cdots \text{ to } m \text{ factors})(a \cdot a \cdots \text{ to } n \text{ factors})$$
$$\equiv (a \cdot a \cdots \text{ to } \overline{m + n} \text{ factors}).$$

Now it has been agreed that an expression,
$$(a \cdot a \cdots \text{ to } m \text{ factors}),$$
shall be symbolized by writing the factor symbol a only once and then writing a number above and at the right to indicate how many such factors there are. By this convention we have
$$(a \cdot a \cdots \text{ to } m \text{ factors}) \equiv a^m;$$
similarly, $\quad (a \cdot a \cdots \text{ to } n \text{ factors}) \equiv a^n.$

Hence, instead of the long form,
$$(a \cdot a \cdots \text{to } m \text{ factors})(a \cdot a \cdots \text{to } n \text{ factors})$$
$$\equiv (a \cdot a \cdots \text{to } m + n \text{ factors}),$$
we have $\qquad (a^m)(a^n) \equiv a^{m+n}.$

The expression a^m is called the mth *power* of a, and m is called the **exponent** because it indicates the power to which a is said to be raised. We may now state the law which is known as the **index law**: *The product of two (or more) quantities consisting of the same quantity raised to different powers is the common quantity with an exponent equal to the sum of the exponents which indicate the powers.*

15. It is important for the student to realize that these primary laws or principles, together with various other laws and theorems yet to be presented, inhere in the nature of quantity and quantitative relations; they are not a device of mathematicians, for they have been discovered, not devised. On the other hand, the system of symbols by means of which these laws are expressed is a device. The use of letters to represent quantity, of the parentheses to indicate aggregation, of the double horizontal bar to express equality — all this is purely conventional. We have an arbitrary shorthand, any feature of which may be discarded at any time for something better if somebody happens to hit upon an improvement. The history of algebra is in part an account of the discovery of laws and theorems, and in part an account of the evolution of the present system of symbols.

16. By an **algebraic expression** is meant an interpretable combination of symbols of quantity, operation, and aggregation. The algebraic expression itself represents a quantity. When part of an expression is connected with other parts by the signs $+$ or $-$, the part is called a **term**.

Expressions are classified as **monomials, binomials, trinomials,** etc., according as they contain one term, two terms, three terms, etc. Binomials, trinomials, quadrinomials, etc., are also conveniently classed together as **polynomials.**

If a term consists of two or more factors and the factors be grouped by any assortment into two groups, one of the groups is the **coefficient,** that is, the co-factor, of the other.

For example, in the expression, $3\,amx$, any grouping may be made of the four factors concerned, and one of the groups becomes the coefficient of the other.

Usually the coefficient group is written first, though this is unnecessary. The context indicates in any case what is to be regarded as the coefficient. Thus in some connections, it may be necessary or desirable to view 3 as the coefficient in the expression $3\,amx$; again, in some other connection it may be $3\,a$, or it may be $3\,am$.

When a term consists of a single letter, unity is to be understood as its coefficient; and similarly unity is its exponent. Thus $x = 1\,x^1$.

<small>The student should guard against the error of saying that a term such as x^n has no coefficient. Such a case never arises. Even if the coefficient is zero, it would be incorrect to say that the term has no coefficient. The same caution holds in regard to exponents.</small>

17. A single letter used as a symbol of quantity is said to be of one **dimension;** the number of dimensions of a term, in which multiplication is the only operation indicated, is the number of letters in the term.

For example, abe, $-b^2e$, $4\,e^3$ are terms of three dimensions; they may also be described as terms of the third

ALGEBRA AS A LANGUAGE.

degree, the degree of a term having reference to the number of its dimensions.

In speaking of the degree of an expression, there may be some reason afforded by the context why a special letter or special letters, rather than all of them, should be referred to as determining the degree.

For example, the expression $-b^2e$ would be described as a term of the first degree in e, and of the second degree in b; as a term in b and e it is of the third degree.

An expression containing at least one term of one dimension and containing no term of more than one dimension is said to be a **linear** expression.

Thus $ax + by$ is a linear expression in x and y; $ax + c$ is a linear expression in x.

The adjective *linear*, used to describe an algebraic feature, is suggestive of geometry. The reason for applying this adjective to a first degree expression will be found in analytic geometry.

In case a term does not contain the quantity which governs the dimensions or degree of an expression, that term is said to be of zero dimensions; it is also called a **constant** or **absolute term**.

In the expression $ax + c$, c is an absolute term, being of the zero degree in x.

An expression is **homogeneous** when it is of the same dimensions in every term.

For example, $ax + by$ is a linear homogeneous expression in x and y; $ax^2 + bx + c$ is a second degree non-homogeneous expression in x; $ax^2 + bxy + cy^2$ is a second degree homogeneous expression in x and y.

18. If an algebraic expression is so constituted that an interchange of two letters leaves the expression unchanged except as to the order of the letters in a term, or as to the order of the terms themselves, the expression is said to be **symmetrical** in the two letters. Since a change in the order of letters in a term, or of terms in an expression, does not alter the value of the quantity which the expression represents, it follows that if an expression is symmetrical in two letters, the expression is not altered in value when the letters are interchanged.

Thus the expression $x^3 + y^3 - 3axy$ is symmetrical in x and y; the expression $ax^2 + bxy + cy^2$ is not symmetrical in x and y unless $a = c$.

CHAPTER II.

ADDITION AND SUBTRACTION.

19. Algebraic addition is the combination of two or more distinct algebraic expressions into one expression by means of the signs $+$ and $-$.

If a and b represent any quantities whatever, positive or negative, their sum is indicated by $a + b$, and we have:

1. the case in which both a and b are positive (the case in ordinary arithmetic);
2. the case in which both a and b are negative;
3. the case in which one quantity is positive and the other negative.

Under the last case, suppose that a is 5 and b is -3; then $a + b$ is $5 + (-3)$; that is, 2. Again, suppose that a is 5 and b is -8; then $a + b$ is $5 + (-8)$; that is, -3.

In general, if we put the sign $-$ before b in order to show that the quantity to be added to a is known to be or is supposed to be negative, we have $a + (-b)$. Now the negative quantity $-b$, combined with the positive quantity a, has the effect of destroying or undoing or neutralizing the quantity $+b$ of the quantity a, so that the result must be expressed: $a - b$. That is, we have $a + (-b) = a - b$.

It thus appears that to add to a positive quantity another quantity viewed as negative, comes to the same thing as to subtract from the positive quantity the second quantity viewed as positive. It follows that one species of addition is subtraction.

It should be noted that no restriction is made as regards the comparative magnitude of a and b. The negative quantity b may be larger than a; in that case $a - b$ is the negative quantity which is left over after all of a has been destroyed or neutralized.

20. The operation of addition is performed as soon as the individual expressions are connected by the signs $+$ and $-$, so that they become terms in the new expression; but usually, by virtue of the *law of distribution* in multiplication, the new expression may be simplified.

Thus the algebraic sum of ax, $-nx$, $4px$, and $-x$ is immediately $ax + (-nx) + 4px + (-x)$; but we notice that x occurs in each one of the given expressions, and that in each expression it is combined with a factor which serves to distinguish the expression from the others and which may be regarded, for the purposes of the required operation, as the coefficient of x. Then by the law of distribution we have

$$ax + (-nx) + 4px + (-x) = (a - n + 4p - 1)x.$$

The advantage of the last expression over the first as regards *form* is that x appears but once, associated with a polynomial coefficient.

We may now state a general rule for addition:

Observe whether any quantity appears as a common factor in each of the expressions to be added; collect by means of their signs all the other factors and write the result as the coefficient of the common quantity. The final expression is the algebraic sum of the given expressions.

ADDITION AND SUBTRACTION.

The coefficient of the new expression will often admit of simplification as regards form; if so, it should of course be simplified. Examples of what is meant by 'simplifying an expression' will be given as we proceed.

21. Since $a - b = a + (-b)$, we have the general rule for subtraction:

Regard the expression to be subtracted as having the opposite sign from that which it really has, and then proceed as in addition.

22. In learning algebra, the student should, from the start, require it of himself to perform mentally every operation which he can. Power will come with practice. It is a mark of faulty training when, at any stage in his progress, the student is found habitually setting down on the blackboard or on paper operations which he ought to "do in his head."

23. Examples illustrative of addition and subtraction.

1. Find the sum of $4b$, $-3b$, $8b$.

The sum is $\quad 4b + (-3b) + 8b$;

but $\quad 4b + (-3b) + 8b = (4 - 3 + 8)b = 9b$.

Such an example as this belongs properly to *mental algebra*. The student should neither write nor say anything except the result: $9b$.

2. Find the sum of a^2, $-a$, a^3.

a is common to the expressions to be added; the corresponding coefficients are a, -1, a^2. They are therefore written in succession, connected by the signs belonging to them, and all enclosed by parentheses. We then have for the sum, $(a - 1 + a^2)a$ But by the *law of commutation* in addition, we may, if we wish, change the order of the terms of the polynomial coefficient. This coefficient will be more orderly, and hence more simple, if the terms are arranged according to their dimensions. We therefore put the result of the addition of a^2, $-a$, a^3, in the form $(a^2 + a - 1)a$.

3. Find the sum of axy, $-bxy$, $(a+b)xy$.

The common quantity is seen to be xy; the coefficients are a, $-b$, $(a+b)$; the sum of these is $a+(-b)+(a+b)$, which is $2a$. The required sum is therefore $2axy$.

4. Write the expression for the sum of each of the following sets of expressions:

1. nx^2y, pxy, qy^2, $-sy$.
2. $(m+n)a$, $-c(m+n)$, $d(m+n)e$.
3. $4(x^2 - xy - y^2)$, $-(-y^2 + x^2 - xy)$, $(-xy - y^2 + x^2)$.

5. Find the sum of
$$a(m+n),\ b(m-n),\ (3+a)(m+n),$$
$$(a+c)(m-n),\ n(m+n).$$

Inspection of these five expressions shows that they have no common factor, but that there is a factor common to two of them and another factor common to the other three. The best that we can do is therefore to form two groups according to the factors $m+n$ and $m-n$. We then have for the sum,
$$(3 + 2a + n)(m+n) + (a+b+c)(m-n).$$

6. Find the sum of
$$ax + by + cz \text{ and } (1-a)x + 2by - \tfrac{2}{3}cz.$$

7. From $nt^2 + t$ subtract $-mt^2 + t$.

8. From $\tfrac{1}{2}x^2 - xy + 2y^2$ subtract $xy + 2y^2 + x^2$.

ADDITION AND SUBTRACTION.

9. From $(a + b)(m - n)$ subtract $(a - b)(m - n)$.

10. Perform the operation indicated by
$$l - (m - [n + \{2a - 3\}]).$$
The parentheses, brackets, and brace are here used to prevent confusion. The brace shows that $2a - 3$ is to be taken as one quantity; therefore we have
$$l - (m - [n + 2a - 3]).$$
Again, the brackets show that $n + 2a - 3$ is to be treated as one quantity, and the sign $-$ preceding the bracketed trinomial indicates that the expression is to be subtracted from m. Similarly, $m - n - 2a + 3$ is to be subtracted from l. Hence we have
$$l - (m - [n + \{2a - 3\}]) = l - (m - [n + 2a - 3])$$
$$= l - (m - n - 2a + 3) = l - m + n + 2a - 3.$$

It will be observed that we began with the inner sign of aggregation and proceeded outwards.

11. For the following expressions write equivalent expressions without symbols of aggregation:

1. $4x - (3ax + [4x - 2ax])$.
2. $2p - (3q + [4q - p])$.
3. $x^2 - cx - (5cx + [1 - 4cx + cx^2])$.

12. Conversely, introduce into parentheses all terms after the first monomial term in the expressions:

1. $\frac{2}{3}px - \frac{1}{2}qx - x^2 + tx^3$.
2. $a - (m + n) - (m - n) + b$.

CHAPTER III.

MULTIPLICATION.

24. Sign of a product. We have already observed that if the quantities a and b are to be multiplied together, the conventional method of indicating it is to write them in juxtaposition, ab. Now if a and b represent any quantities whatever, positive or negative, there must arise three cases just as in addition:

1. both quantities positive;
2. both quantities negative;
3. one quantity positive and the other negative.

We have to inquire what sign belongs to the product ab in each case.

The first case is that of ordinary arithmetic and requires no special attention; if a and b are each positive, their product is positive.

To meet the other cases, we notice that a negative quantity comes by taking a positive quantity once and then changing its sign. To take it once is to multiply it by unity, and by means of the symbol -1 we may show the two operations: multiplying by unity, and reversing the sign.

Thus, $\quad -b = [-][1][b] = [-1][b];$
hence, $\quad [a][-b] = [-1][b][a].$

By this symbolism we mean that the positive quantity a is multiplied by the positive quantity b, resulting in

MULTIPLICATION. 21

the positive quantity ab; ab is taken once and its sign then reversed, as indicated by the symbol $[-1]$.

Hence, $(a)(-b) = -ab$.

Again, $[-a][-b] = [b][a][-1][-1]$.

Proceeding from left to right and performing the operations in the order indicated, we have first the positive quantity ab; the third bracketed symbol shows that ab is now to be taken once and given the negative sign, thus becoming $-ab$; finally the last symbol shows that the negative quantity ab is to be taken once and given the opposite sign.

Hence we state briefly: *Like signs give $+$ and unlike signs $-$*.

Applying the argument used to show that $(-a)(-b)$ is $+ab$, we may also show that $(-a)(-b)(-c)$ is $-abc$. Continuing this operation, we have the general conclusion: *An even number of negative quantities multiplied together is a positive quantity, whilst an odd number of negative quantities multiplied together is a negative quantity.*

25. Product of two monomials. If an expression containing two or more factor symbols is to be multiplied by another expression containing two or more factor symbols, the product is indicated by writing all the factor symbols in juxtaposition. Thus,

$$(amx)(bny) = amxbny.$$

By the *law of commutation* the order of these factors does not affect the value of the result. Further, as seen above, the sign of the product is determined by the number of negative factors.

Some of the factor symbols may occur in each of the expressions to be multiplied together; but we have noticed that the conventional way of indicating the product of n equal factors is to write the factor symbol only once and then write in connection with it the exponent n.

For example,
$$(amx)(anx^2) = amxanxx = a^2mnx^3.$$

We may now frame the following rule for the multiplication of one monomial by another:

Write in juxtaposition in whatever order seems desirable all the factors of the two monomials. If any one factor occurs more than once, write it only once and give it an exponent equal to the sum of its exponents in the two monomials. The result will be positive or negative according as there is an even or an odd number of negative factors.

26. Product of a monomial and a polynomial. By the *law of distribution*,
$$(a + b)c = ac + bc,$$
whatever may be the values of a, b, and c; it must therefore be true if we regard a as made up of m and n.

Substituting $m + n$ for a,
$$[(m+n)+b]c = (m+n)c + bc = mc + nc + bc;$$
that is, $\quad (m+n+b)c = mc + nc + bc.$

If besides substituting $m + n$ for a, we had supposed $p - q = b$ and had substituted $p - q$ for b, we should have obtained
$$(m + n + p - q)c = mc + nc + pc - qc;$$
and evidently the result may be extended to the multi-

MULTIPLICATION. 23

plication of a polynomial of any number of terms by a monomial.

Hence we have the rule:

Multiply each term of the polynomial, taken with the sign which precedes the term, by the monomial; the algebraic sum of these part products is the product of the two expressions.

27. Product of two polynomials. In the statement,
$$(m + n + p - q)c = mc + nc + pc - qc,$$
suppose $c = x + y;$
then we have
$$(m + n + p - q)(x + y)$$
$$= m(x + y) + n(x + y) + p(x + y) - q(x + y)$$
$$= mx + my + nx + ny + px + py - qx - qy;$$
or, re-arranging the terms with respect to x,
$$= mx + nx + px - qx + my + ny + py - qy.$$

Evidently the case can be extended so as to be even more general, and we have a rule for the product of a polynomial of any number of terms multiplied by any other polynomial:

Multiply each term of the multiplicand, taken with the sign which precedes it, by each term of the multiplier, taken with the sign which precedes it; the algebraic sum of these part products is the product of the two polynomials.

28. As has already been stated, if a quantity occurs twice as a factor, that is if it is multiplied by itself, we write the quantity once and give it an exponent 2; if a quantity is multiplied by itself twice, we write it once with an exponent 3; and if a quantity is multiplied by itself $n-1$ times, we write it once with an exponent n.

The term **power** is applied to such expressions as x^2, x^3, x^n; and we speak of the *second power of x*, the nth *power of x*, etc.

The second power of a quantity is also called the '*square*' of that quantity, and the third power is called the '*cube*.'

Thus x^2 is read 'x *squared*'; x^3 is read 'x *cubed*.'

29. The square of the sum of two quantities is of such frequent occurrence that it is important to memorize the expression which results when the operation indicated by the exponent is actually performed.

Let a and b be any two quantities; then $(a+b)^2$ is an *indicated*, that is, an *unperformed* operation, and means that we have $(a+b)$ multiplied by itself.

$$\text{Now } (a+b)^2 = (a+b)(a+b) = a(a+b) + b(a+b)$$
$$= aa + ab + ba + bb$$
$$= a^2 + 2ab + b^2.$$

"It will be instructive to write out this shorthand at length. The square of the sum of two numbers means that sum multiplied by itself. But this product is the first number multiplied by the sum together with the second number multiplied by the sum. Now the first number multiplied by the sum is the same as the first number multiplied by itself together with the first number multiplied by the second number. And the second number multiplied by the sum is the same as the second number multiplied by the first number together with the second number multiplied by itself. Putting all these together, we find that the square of the sum is equal to the sum of the squares of the two numbers, together with twice their product.

Two things may be observed on this comparison. First, how very much the shorthand expression gains in clearness from its brevity. Secondly, that it is only shorthand for something which is just straightforward common sense and nothing else. We may always depend upon it that algebra which cannot be translated into good English and sound common sense is bad algebra." *

Omitting the steps in the proof and corresponding portions of the translation, the statement

$$(a + b)^2 = a^2 + 2\,ab + b^2$$

affords the theorem,

The square of the sum of any two quantities is equal to the sum of their squares plus twice their product.

30. Similarly, we have

$$(a - b)^2 = (a - b)(a - b)$$
$$= aa + (-b)a + a(-b) + (-b)(-b)$$
$$= a^2 - 2\,ab + b^2.$$

Hence, *the square of the difference of any two quantities is equal to the sum of their squares minus twice their product.*

It will be noticed that we might have obtained this theorem even more directly by writing $-b$ for b in the first formula. For we have

$$(a - b)^2 = [a + (-b)]^2 = a^2 + 2\,a(-b) + (-b)^2$$
$$= a^2 - 2\,ab + b^2.$$

---

* Clifford's *The Common Sense of the Exact Sciences.*

31. To find the product of the sum and the difference of any two quantities, we have

$$(a+b)(a-b) = aa + ab + (-b)a + (-b)b$$
$$= a^2 + ab - ab - b^2$$
$$= a^2 - b^2.$$

That is, *the product of the sum and the difference of any two quantities is equal to the difference of their squares.*

32. The results in Arts. 29, 30, 31 afford good examples of truths, more or less general, expressed in the language of algebra and called *formulas*, the translations of which into ordinary language are called *theorems*.

It will be a valuable exercise for the student to prove the theorems of Arts. 29, 30, 31 by the *graphic* method.

Thus, to prove the first theorem, let a straight line AB represent a in magnitude. Produce AB to C, making BC represent b in magnitude. Construct squares on AB and AC, and notice the areas corresponding to the terms of the trinomial $a^2 + 2ab + b^2$.

33. Product of homogeneous polynomials. The product of any two homogeneous expressions is homogeneous; for if each term of one of the expressions is of m dimensions, and each term of the other is of n dimensions, each term of the product must be of $m+n$ dimensions, since it is obtained by multiplying a term of the multiplicand by a term of the multiplier.

EXAMPLES.

1.. Multiply $x^2 - \frac{2}{3}xy + 2y^2$ by $\frac{3}{4}x^2 - \frac{1}{4}xy + \frac{1}{4}y^2$.

2. Multiply together $a+b+c$, $-a+b+c$, $a-b+c$, $a+b-c$.

MULTIPLICATION.

3. Derive a formula for the cube of the sum of two quantities, and translate the formula into a theorem.

4. By means of the formula for $(a+b)^3$, find the cube of $2x+5y$. Verify the result by performing the indicated operation $(2x+5y)^3$ without the aid of the formula.

5. Find $(a-b)^3$.

6. By means of the result in Ex. 5, find $(\tfrac{1}{2}x - \tfrac{2}{3}y)^3$.

7. Find $(a+b+c)^3$. Consider whether the result is homogeneous in a, b, and c.

8. Show how $(a+b+c)^2$ may be found by treating $a+b+c$ as a binomial.

9. Show that if two linear expressions are multiplied together, the product must contain a term of two dimensions as regards the quantities which made the given expressions linear.

10. Show that
$$(a+b)^2 + 2c(a+b) + b^2 = (b+c)^2 + 2a(b+c) + a^2$$
$$= (a+c)^2 + 2b(a+c) + b^2.$$
Show that each one of these expressions is homogeneous in a, b, and c. Also, that each one of them is symmetrical with respect to any two of the three letters in it.

11. Prove that
$$p^2 + (p+q)^2 - p(p+q) = q^2 + (q+p)^2 - q(q+p).$$
Is the expression $p^2 + (p+q)^2 - p(p+q)$ homogeneous in p and q? Is it symmetrical with respect to p and q? If p and q have exchanged places, and the new expression equals the old one, why does it not follow that the expression is symmetrical as regards p and q?

CHAPTER IV.

DIVISION

34. In division a quantity, the dividend, is given together with another quantity, the divisor. The divisor may be regarded as a factor of the first quantity; it is then required to find the other factor, the quotient. Division is thus the reverse of multiplication. Hence, various rules to be observed in performing division may be inferred at once from the rules for multiplication.

35. Division of any power of a quantity by any other power of that quantity.

Let a^m represent $a \cdot a \cdots$ to m factors; that is, the mth power of any quantity a. Also, let a^n represent the nth power of a.

Two cases arise, according as $m > n$ or $m < n$, with a transition case when $m = n$.

1st. Suppose that $m > n$.

We have $\quad \dfrac{a^m}{a^n} = \dfrac{a \cdot a \cdot a \cdots \text{ to } m \text{ factors}}{a \cdot a \cdot a \cdots \text{ to } n \text{ factors}}.$

Since, by the *law of association*, factors may be associated in any manner whatever, we may place in one group n of the m factors of the dividend, and in another group the remaining $\overline{m-n}$ factors, so that we have

$$\dfrac{a^m}{a^n} = \dfrac{(a \cdot a \cdot a \cdots \text{ to } n \text{ factors})(a \cdot a \cdot a \cdots \text{ to } \overline{m-n} \text{ factors})}{(a \cdot a \cdot a \text{ to } n \text{ factors})}.$$

DIVISION.

Now by the law just quoted, $pqr = p(qr)$; hence $\dfrac{pqr}{p} = \dfrac{p}{p}(qr)$, which is qr. That is, one quantity may be divided by another by dividing one *factor* of the first quantity by the second quantity. Therefore, dividing the first group of a's in the dividend of the above expression by the divisor, we have

$$\frac{a^m}{a^n} = (a \cdot a \cdot a \cdots \text{to } \overline{m-n} \text{ factors}) = a^{m-n}.$$

It follows, therefore, that *the quotient of any power of a quantity divided by a lower power of that quantity is the quantity raised to a power indicated by the exponent of the dividend minus the exponent of the divisor.*

2d. Suppose that $m < n$.

As before, we have

$$\frac{a^m}{a^n} = \frac{(a \cdot a \cdot a \cdots \text{to } m \text{ factors})}{(a \cdot a \cdot a \cdots \text{to } m \text{ factors})(a \cdot a \cdot a \cdots \text{to } \overline{n-m} \text{ factors})},$$

in which the n factors of the divisor are grouped into m factors and $\overline{n-m}$ factors; and again, since $pqr = p(qr)$, $\dfrac{p}{pqr} = \dfrac{p}{p(qr)} = \dfrac{1}{qr}$. That is, one quantity may be divided by another by dividing the first quantity by one *factor* of the second quantity.

Hence we have $\dfrac{a^m}{a^n} = \dfrac{1}{a^{n-m}}$,

when $n > m$; that is, when any power of a quantity is divided by a higher power of a quantity.

It will now be convenient to adopt the convention $\dfrac{1}{a^{n-m}} = a^{m-n}$, for, by so doing, a rule may be stated which will cover the first case and the second case also:

The quotient of any power of a quantity divided by any other power of that quantity is the quantity with an exponent equal to the exponent of the dividend minus the exponent of the divisor.

When $n > m$, in $\dfrac{a^m}{a^n}$, $m - n$ is of course negative. It appears, therefore, that negative integral exponents originate in division when a power of a quantity is divided by a higher power of the quantity; and it must be carefully remembered that any quantity with such an exponent is the same as the reciprocal of that quantity with the exponent taken as positive

If $m = n$, $\dfrac{a^m}{a^n}$ becomes $\dfrac{a^m}{a^m}$,

and by the above law,

$$\frac{a^m}{a^m} = a^{m-m} = a^0.$$

But any quantity divided by itself is unity; hence

$$\frac{a^m}{a^m} = 1;$$

therefore $\qquad a^0 = 1;$

that is, *any quantity with zero for an exponent is unity.*

36. Quotient of one monomial divided by another. The rule is obtained at once from the corresponding rule in multiplication:

Compare the factor symbols of the divisor with those of the dividend.

Write down in any desired order such of the factor symbols as are found in the dividend and not in the divisor.

DIVISION.   31

Also, write down, after changing the sign of its exponent, any factor found in the divisor and not in the dividend.

If a factor is found common to both dividend and divisor, write its symbol once in the quotient with an exponent equal to its exponent in the dividend minus its exponent in the divisor.

Give to the quotient the sign $+$ if the dividend and divisor have like signs; otherwise give it the sign $-$.

It is left to the student to establish this rule by considering that division is the reverse of multiplication, and that the correctness of an operation in division is proved by multiplication.

37. Quotient of a polynomial divided by a monomial. Referring to the corresponding rule in multiplication, we have the rule:

Divide each term of the polynomial, taken with its sign, by the monomial. The algebraic sum of the part quotients is the required quotient.

For example, let it be required to divide
$$-4c^2x^3 + ac^{-1}x^2 - \tfrac{1}{2}x + 10c \text{ by } 2cx.$$

The operation is as follows:
$$2cx)\; \underline{-4c^2x^3 + ac^{-1}x^2 - \tfrac{1}{2}x + 10c}$$
$$-2cx^2 + \tfrac{1}{2}ac^{-2}x - \tfrac{1}{4}c^{-1} + 5x^{-1}.$$

38. Quotient of one polynomial divided by another. This is the most general case of division, though not the most common. An example will illustrate the process and will conduct to a rule. Suppose it is required to divide
$$x^3 - 3axy + y^3 + a^3 \text{ by } x + y + a.$$

The complete operation is as follows:

$$x^3 - 3axy + y^3 + a^3 \, (x + y + a$$

(1) ... $x^3 + x^2y + ax^2 \qquad (x^2 - xy + y^2 - ax - ay + a^2$

$\overline{-x^2y - ax^2 - 3axy + y^3 + a^3}$

(2) . $-x^2y - xy^2 - axy$

$\overline{xy^2 - ax^2 - 2axy + y^3 + a^3}$

(3) .. $xy^2 + y^3 + ay^2$

$\overline{- ax^2 - 2axy - ay^2 + a^3}$

(4) .... $-ax^2 - axy - a^2x$

$\overline{ - axy - ay^2 + a^2x + a^3}$

(5) .... $-axy - ay^2 - a^2y$

$\overline{ a^2x + a^2y + a^3}$

(6) .... $a^2x + a^2y + a^3$

For convenience in multiplication the divisor was written at the right of the dividend. As soon as any term of the quotient was found, it was written in order beneath the divisor.

x^3, the first term of the dividend, was divided by x, the first term of the divisor; and x^2, the quotient, was written for the first term of the required quotient.

The entire divisor was then multiplied by the first term of the quotient; and the product, numbered (1), was subtracted from the dividend.

$-x^2y$, the first term of the remainder, was then divided by the first term of the divisor; and $-xy$, the quotient, was written as the second term of the quotient.

As before, the entire divisor was multiplied by this term of the quotient, and the product (2) was subtracted from the first remainder.

In this manner the process was carried on until there was no remainder.

What has practically been done is to break up the given polynomial dividend into part polynomials, numbered (1), (2), (3), (4), (5), (6), and divide each one of these parts by the given divisor. The part polynomial dividends are seen to be so constituted that each could be exactly divided by the divisor with a monomial quotient, the algebraic sum of these monomial quotients becoming the final required quotient.

That the given dividend has been thus broken up into part dividends will be seen by adding together (1), (2), (3), (4), (5), (6), thus:

$$\begin{aligned}
(1) &\quad x^3 + x^2y + ax^2 \\
(2) &\quad - x^2y - xy^2 - axy \\
(3) &\quad xy^2 + y^3 + ay^2 \\
(4) &\quad - ax^2 - axy - a^2x \\
(5) &\quad - axy - ay^2 - a^2y \\
(6) &\quad a^2x + a^2y + a^3 \\
\hline
&\quad x^3 - 3axy + y^3 + a^3
\end{aligned}$$

The reasons for the various steps taken in the solution of the above example hold for any similar example. Hence we have the rule:

Arrange both dividend and divisor according to the ascending or descending powers of some letter common to both.

Divide the first term of the dividend by the first term of the divisor; this will give the first term of the quotient.

Multiply the whole divisor by the first term of the quotient as just found, and subtract the product from the dividend.

Repeat the process until there is no remainder, or until the remainder is of a lower degree than the divisor.

EXAMPLES.

1. Divide $-7a^2m^8x$ by $4am^4$.
2. Divide $a^3 - b^3$ by $a - b$.
3. Divide $x^3 + a^3$ by $x + a$.
4. Divide $x^4 - a^4$ by $x - a$. What is the degree of the quotient as regards x? Is it homogeneous as regards x?
5. Divide $x^3 + y^3 - 3axy + a^3$ by $x + y + a$ by treating $x + y + a$ as a binomial $(x + y) + a$ and using the result of Ex. 3.
6. If $x^2 + 3x + a$ is exactly divisible by $x + 2$, what is the value of a?

CHAPTER V.

THEORY OF EXPONENTS.

39. Integral exponents. We have seen in previous chapters that
$$(a^m)(a^n) = a^{m+n},$$
and
$$\frac{a^m}{a^n} = a^{m-n},$$

m and n representing integral numbers.

It may also be shown that
$$(a^m)^n = a^{mn},$$
$$(ab)^m = a^m b^m,$$
$$\left(\frac{a}{b}\right)^m = \frac{a^m}{b^m},$$

m and n being positive and integral.

For $(a^m)^n = a^m \cdot a^m \cdot a^m \cdots$ to n factors,
$$= a^{m+m+m\cdots \text{ to } n \text{ terms}},$$
$$= a^{nm} = a^{mn}.$$

Similarly $(ab)^m = ab \cdot ab \cdot ab \cdots$ to m factors,
$= (a \cdot a \cdot a \cdots \text{ to } m \text{ factors})(b \cdot b \cdot b \cdots \text{ to } m \text{ factors}),$
$= a^m b^m.$

36   ALGEBRA.

Also $\quad \left(\dfrac{a}{b}\right)^m = \dfrac{a}{b} \cdot \dfrac{a}{b} \cdot \dfrac{a}{b} \cdots$ to m factors,

$$= \dfrac{a \cdot a \cdot a \cdots \text{ to } m \text{ factors}}{b \cdot b \cdot b \cdots \text{ to } m \text{ factors}},$$

$$= \dfrac{a^m}{b^m}.$$

Thus we have the five formulas:

1. $\quad (a^m)(a^n) = a^{m+n}.$
2. $\quad \dfrac{a^m}{a^n} = a^{m-n}.$
3. $\quad (a^m)^n = a^{mn}.$
4. $\quad (ab)^m = a^m b^m.$
5. $\quad \left(\dfrac{a}{b}\right)^m = \dfrac{a^m}{b^m}.$

It remains to note the origin and significance of *fractional* exponents.

40. Reciprocal exponents. Suppose that any quantity a is to be resolved into m equal factors. Can we give a an exponent such that if a with this exponent is raised to the mth power the result shall be simply a? If this can be done, the exponent in question serves to indicate one of the m equal factors of a, or the mth **root** of a, as it is usually called.

Let us write

$$a = a^{(\ )} \cdot a^{(\ )} \cdot a^{(\ )} \cdots \text{ to } m \text{ factors.}$$

What common exponent is required within the parentheses?

By the law for multiplication, we have from the assumed expression

THEORY OF EXPONENTS.

$$a^{(\,)} \cdot a^{(\,)} \cdot a^{(\,)} \cdots \text{ to } m \text{ factors}$$
$$= a^{(\,)+(\,)+(\,) \text{ to } m \text{ terms}}$$
$$= a^{m(\,)};$$

hence $\quad a = a^{m(\,)};$

but a means a^1;

therefore $\quad m(\,) = 1,$

and hence the quantity required within the parentheses can be none other than $\dfrac{1}{m}$; and we have $a^{\frac{1}{m}}$, m being positive and integral, signifying the mth root of a.

•For example, the *square root* of any quantity is indicated by giving the quantity the exponent $\tfrac{1}{2}$; the *cube root*, that is, one of the three equal factors of a quantity, is indicated by the exponent $\tfrac{1}{3}$.

The student should carefully observe that this meaning of an exponent of the form $\dfrac{1}{m}$, m being positive and integral, is not itself a convention, it follows necessarily from the primary convention that the product of m factors, each equal to a, shall be indicated by a^m.

41. Positive fractional exponents. Suppose that any quantity, as a, is resolved into m equal factors and n of these factors are then multiplied together; we have thus a power of a root, and it is evidently expressed by $\left(a^{\frac{1}{m}}\right)^n$.

But $\quad \left(a^{\frac{1}{m}}\right)^n = a^{\frac{1}{m}} \cdot a^{\frac{1}{m}} \cdot a^{\frac{1}{m}} \cdots \text{ to } n \text{ factors}$
$$= a^{\frac{1}{m}+\frac{1}{m}+\frac{1}{m}+ \cdots \text{ to } n \text{ terms}}$$
$$= a^{\frac{n}{m}}.$$

Therefore, *to raise any root of a quantity to any power, multiply the exponent indicating the root by the index of the power*

It has just been shown that a positive fractional exponent indicates two operations. If we can show that

$$\left(a^{\frac{1}{m}}\right)^n = (a^n)^{\frac{1}{m}},$$

we must conclude that the order in which these operations are performed does not affect the result.

Writing as in the preceding article,

$$a^n = a^{n(\ )} \cdot a^{n(\ )} \cdot a^{n(\ )} \cdots \text{ to } m \text{ factors,}$$

we have
$$a^n = a^{n(\ )+n(\ )+n(\ )+\ \text{to } m \text{ terms}}$$
$$= a^{mn(\ )};$$

hence
$$mn(\ ) = n,$$

and therefore the quantity required within the parentheses must be $\frac{1}{m}$, and we have $a^{n\left(\frac{1}{m}\right)}$ as the mth root of a^n;

but
$$a^{n\left(\frac{1}{m}\right)} = a^{\frac{n}{m}},$$

and as $\left(a^{\frac{1}{m}}\right)^n$ has also been found equal to $a^{\frac{n}{m}}$, it follows that

$$\left(a^{\frac{1}{m}}\right)^n = (a^n)^{\frac{1}{m}}.$$

But the first of these expressions indicates a power of a root, and the second indicates a root of a power. Therefore the order of performing these operations does not affect the result.

42. Negative fractional exponents. A quantity affected with a negative fractional exponent is equal to the reciprocal of the quantity with the sign of the exponent changed.

For suppose we have

$$\frac{a^{\frac{n}{m}}}{a^{\frac{p}{q}}}, \text{ in which } \frac{n}{m} < \frac{p}{q},$$

that is, $\quad\dfrac{nq}{mq} < \dfrac{mp}{mq},$

and therefore $\quad nq < mp;$

(see chapter on inequalities).

Now let $\quad a^{\frac{1}{mq}} = b;$

then $\quad \dfrac{a^{\frac{n}{m}}}{a^{\frac{p}{q}}} = \dfrac{a^{\frac{nq}{mq}}}{a^{\frac{mp}{mq}}} = \dfrac{b^{nq}}{b^{mp}} = b^{nq-mp} = \dfrac{1}{b^{-(nq-mp)}}.$

Replacing b by its value $a^{\frac{1}{mq}}$,

$$\left(a^{\frac{1}{mq}}\right)^{(nq-mp)} = \frac{1}{\left(a^{\frac{1}{mq}}\right)^{-(nq-mp)}};$$

that is, $\quad a^{\frac{n}{m}-\frac{p}{q}} = \dfrac{1}{a^{-\left(\frac{n}{m}-\frac{p}{q}\right)}}.$

43. Employing the method of Art. 42, we may show that

$$\left(a^{\frac{n}{m}}\right)\left(a^{\frac{p}{q}}\right) = a^{\frac{n}{m}+\frac{p}{q}}.$$

Formulas 1 and 2 of Art. 39 are thus found to hold for fractional exponents as well as for integral.

To prove formula 3 for fractional exponents, that is, to show that

$$\left(a^{\frac{n}{m}}\right)^{\frac{p}{q}} = a^{\frac{np}{mq}},$$

we notice that it has already been shown (Art. 41) that
$$\left(a^{\frac{n}{m}}\right)^p = a^{\frac{pn}{m}};$$

consequently $\left(a^{\frac{n}{m}}\right)^{\frac{p}{q}} = \left(a^{\frac{pn}{m}}\right)^{\frac{1}{q}},$

and we have only to prove that the qth root of this last expression is indicated by multiplying the denominator by q.

As in former cases, let

$$a^{\frac{pn}{m}} = a^{\frac{pn}{m}(\ )} \cdot a^{\frac{pn}{m}(\ )} \cdot a^{\frac{pn}{m}(\ )} \cdots \text{ to } q \text{ factors}$$
$$= a^{\frac{pn}{m}(\ )+\frac{pn}{m}(\ )+\cdots\text{ to } q \text{ terms}}$$
$$= a^{\frac{qpn}{m}(\ )}.$$

Therefore $\dfrac{pn}{m} = q\dfrac{pn}{m}(\ ),$

and hence the parenthetical factor must be $\dfrac{1}{q}$.

Hence one of the q equal factors of $a^{\frac{pn}{m}}$ is $a^{\frac{pn}{qm}}$;

therefore $\left(a^{\frac{n}{m}}\right)^{\frac{p}{q}} = a^{\frac{np}{mq}}.$

44. To show that $(ab)^{\frac{n}{m}} = a^{\frac{n}{m}}b^{\frac{n}{m}},$
we have already $(ab)^n = a^n b^n;$

therefore $(ab)^{\frac{n}{m}} = (a^n b^n)^{\frac{1}{m}}$
$= [(a \cdot a \cdot a \cdots \text{to } n \text{ factors})(b \cdot b \cdot b \cdots \text{to } n \text{ factors})]^{\frac{1}{m}}.$

Since factors may be grouped in any manner without affecting the result, let us group the n factors of the first

THEORY OF EXPONENTS.

parenthetical group into m equal groups, and the n factors of the second parenthetical group into m equal groups. If we now associate one of the first of these factors with one of the second, we shall evidently have one of the m equal factors of the expression within the brackets; but a factor out of the first group is $a^{\frac{n}{m}}$ and one out of the second is $b^{\frac{n}{m}}$.

Hence $$(a^n b^n)^{\frac{1}{m}} = a^{\frac{n}{m}} b^{\frac{n}{m}};$$

and therefore $$(ab)^{\frac{n}{m}} = a^{\frac{n}{m}} b^{\frac{n}{m}}.$$

45. Finally, we may show that

$$\left(\frac{a}{b}\right)^{\frac{n}{m}} = \frac{a^{\frac{n}{m}}}{b^{\frac{n}{m}}},$$

by making it a case under the formula just established.

For $$\frac{a}{b} = a\left(\frac{1}{b}\right) = ab^{-1},$$

and hence $$\left(\frac{a}{b}\right)^{\frac{n}{m}} = (ab^{-1})^{\frac{n}{m}} = a^{\frac{n}{m}} (b^{-1})^{\frac{n}{m}};$$

but $$(b^{-1})^{\frac{n}{m}} = \left(b^{\frac{n}{m}}\right)^{-1} \qquad \text{(Art. 41)}$$

$$= \frac{1}{b^{\frac{n}{m}}};$$

therefore $$\left(\frac{a}{b}\right)^{\frac{n}{m}} = a^{\frac{n}{m}}\left(\frac{1}{b^{\frac{n}{m}}}\right) = \frac{a^{\frac{n}{m}}}{b^{\frac{n}{m}}}.$$

46. Since formulas 1, 2, 3, 4, 5 of Art. 39 have been shown to hold for fractional exponents as well as for integral exponents, the translation of the formulas must accord with this fact.

For example, formula 5 is equivalent to the theorem:

A quotient affected with any exponent, integral or fractional, is equal to the quotient obtained if the dividend and divisor are first separately affected with the exponent.

EXAMPLES.

1. Establish the statement:

$$\frac{a^m}{a^n} = a^{m-n},$$

(1) when m is positive and n negative,
(2) when m is negative and n positive,
(3) when m is negative and n is negative.

2. Simplify the following expressions:

$$[(x^4)b^{-\frac{2}{3}}]^{\frac{3}{2}}; \qquad (a^{\frac{3}{2}}c^{-\frac{1}{3}})^{\frac{2}{3}};$$

$$x^{\frac{3}{2}} \cdot x^{-\frac{2}{3}} \cdot x^{-1}; \qquad (y^{\frac{1}{2}})^2 + y^0.$$

3. What are the numerical values of the following expressions:

$$8^{\frac{1}{3}} \div (-2)^3; \qquad (100)^{\frac{1}{2}}(\tfrac{1}{16})^{-\frac{1}{4}};$$

$$(\tfrac{1}{36})^{-\frac{1}{2}}(36)^{\frac{1}{2}}; \qquad 8^{-\frac{1}{3}}(7^2)^{-\frac{1}{2}}.$$

4. Write the square of $\dfrac{a}{2}\left(e^{\frac{x}{a}} + e^{-\frac{x}{a}}\right)$.

5. Multiply $a^{2n} - (1)^0 + a^n$ by $a^{-2n} + a^{-n} + (1)^{-1}$.

6. Show by the use of negative exponents that

$$\left(\frac{a}{b}\right)\left(\frac{m}{n}\right) = \frac{am}{bn};$$

and thence state a rule for the multiplication of one fraction by another.

7. Show in a similar manner that

$$\frac{a}{b} \div \frac{m}{n} = \left(\frac{a}{b}\right)\left(\frac{n}{m}\right).$$

8. Examine the argument of Art. 41,

if
$$a^n = a^{mn(\ )},$$
$$mn(\ ) = n.$$

9. If $x^p = y^q$ and $p = q$, $x = y$. Prove and translate this statement.

CHAPTER VI.

SURDS.

47. In the preceding chapter it has been shown that the exponent $\dfrac{1}{m}$, m being positive and integral, denotes that the quantity to which it belongs is to be resolved into m equal factors and one of the factors taken. This factor or mth root is also indicated by the **radical sign** $\sqrt[m]{}$, and we have by definition $\sqrt[m]{a}$ is $a^{\frac{1}{m}}$, m being the **index** of the root. If m is 2 in any case, it is customary not to express the index.

Thus $a^{\frac{1}{2}}$ is also written $\sqrt{a}$.

48. A **surd** is an indicated root which cannot be exactly extracted.

Thus $\sqrt{5}$, $\sqrt[5]{a^2}$, $\sqrt{x^2+y^2}$ are surds.

The term **irrational quantity** is also applied to a quantity such that an indicated root cannot be exactly found.

Quantities not belonging to the class *irrational* are termed **rational**.

The radical sign may occur in connection with a rational quantity; that is, we may have rational quantities in surd form.

$\sqrt{4x^2}$, $\sqrt[3]{8}$, $\sqrt[m]{a^{2m}}$ are examples of rational quantities in surd form.

Both true surds and pseudo-surds are often called

SURDS. 45

radicals. We shall use the terms *surd* and *radical* interchangeably.

The **order** of a surd is indicated by the surd index or root symbol.

Thus $\sqrt{a}$ is a surd of the second order; $\sqrt[m]{b}$ is a surd of the mth order.

Surds of the second order are also called **quadratic surds.**

A **mixed surd** is one which can be resolved into an **irrational factor** and a **rational factor.**

The rational factor is to be viewed as a coefficient of the true surd.

When the irrational factor is integral, the surd is in its simplest form.

Also, when surds of the same order contain the same irrational factor, they are said to be **similar** or **like.**

49. Since a surd can always be expressed by means of a fractional exponent, it is evident that rules for the treatment of surds must be sought in the principles of exponents.

For example, if we wish to simplify the third order surd $\sqrt[3]{16\,a^4 x^6}$, we notice that

$$\sqrt[3]{16\,a^4 x^6} = (16\,a^4 x^6)^{\frac{1}{3}} = (8)^{\frac{1}{3}}(2)^{\frac{1}{3}}(a^3)^{\frac{1}{3}}(a)^{\frac{1}{3}}(x^6)^{\frac{1}{3}}.$$

Performing the operations indicated, so far as they can be performed,

$$\sqrt[3]{16\,a^4 x^6} = 2\,ax^2 (2\,a)^{\frac{1}{3}} = 2\,ax^2 \sqrt[3]{2\,a}.$$

It thus appears that the given expression is a mixed surd, and that by means of theorems regarding exponents we are able to resolve it into an irrational factor $\sqrt[3]{2\,a}$

and a rational one $\sqrt[3]{8\,a^3x^6}$; this last factor is equal to $2\,ax^2$ and becomes the coefficient of the first factor.

The student is not to infer from this example, however, that in order to handle a surd it is necessary to convert it into an equivalent expression with fractional exponents. In practice, surds (radicals) are usually transformed without the introduction of fractional exponents.

50. Reduction of a surd to its simplest form. Let ab be any quantity whose mth root is indicated.

Then
$$\sqrt[m]{ab} = (ab)^{\frac{1}{m}} = a^{\frac{1}{m}}b^{\frac{1}{m}} \qquad \text{(Art. 44)}$$
$$= \sqrt[m]{a}\sqrt[m]{b}.$$

If now a and m are such quantities that we can actually extract the mth root of a, whilst b is such a quantity that it contains no factor whose mth root can be extracted, the given surd is reduced to its simplest form by expressing it $\sqrt[m]{a}\sqrt[m]{b}$.

This process may evidently be applied to a quantity containing any number of factors.

51. Surds of different orders can be reduced to the same order; for if we have $\sqrt[m]{a}$ and $\sqrt[n]{b}$,

$$\sqrt[m]{a} = a^{\frac{1}{m}} = a^{\frac{n}{nm}} = \sqrt[nm]{a^n},$$
and
$$\sqrt[n]{b} = b^{\frac{1}{n}} = b^{\frac{m}{mn}} = \sqrt[nm]{b^m}.$$

These surds are reduced to the same order by raising each quantity under the radical sign to a power equal to the surd index of the other quantity, and then giving to each resulting quantity a surd index equal to the product of the two surd indices.

If the surd indices have a least (or lowest) common multiple, we may take advantage of the fact.

For example, let it be required to reduce $\sqrt[6]{a}$ and $\sqrt[8]{b}$ to the same order. 24 is seen to be the least common multiple of the indices.

$$\sqrt[6]{a} = a^{\frac{1}{6}} = a^{\frac{4}{24}} = \sqrt[24]{a^4};$$

and $$\sqrt[8]{b} = b^{\frac{1}{8}} = b^{\frac{3}{24}} = \sqrt[24]{b^3}.$$

52. Addition of surds. *Reduce the given surds to like surds of the simplest form. Take the algebraic sum of the coefficients of the common surd factor, and write this sum as the coefficient of the surd factor.*

If surds are unlike, whether the unlikeness is due to the surds being of different degrees, or being different irrational quantities of the same degree, their coefficients cannot be collected.

53. Subtraction of surds. *Reduce the two given surds to like surds of the simplest form. Regard the sign of the subtrahend as the opposite of that which it actually is, and proceed as in addition.*

54. Product of two surds of the same order. If it is required to multiply $p\sqrt[m]{a}$ by $q\sqrt[m]{b}$, we have

$$(p\sqrt[m]{a})(q\sqrt[m]{b}) = pq\left(a^{\frac{1}{m}}\right)\left(b^{\frac{1}{m}}\right) = pq\,(ab)^{\frac{1}{m}} = pq\sqrt[m]{ab};$$

hence, to multiply together two (or more) surds of the same order, *multiply together the quantities under the radical sign, and write, as a coefficient of the surd, the product of the rational factors.*

To multiply together surds of different orders, *reduce the given surds to equivalent surds of the same order, and proceed as before.*

55. Quotient of two surds of the same order. The rule is inferred at once from the corresponding rule for multiplication.

Divide the rational factor of the dividend by the rational factor of the divisor; make this quotient the coefficient of the irrational factor of the dividend divided by the irrational factor of the divisor.

Similarly, to divide a surd of one order by a surd of another order, *reduce them to equivalent surds of the same order and proceed as in the previous case.*

56. Rationalization of the denominator. If the division of one surd by another is indicated by the horizontal bar, giving a surd fraction of the form $\dfrac{\sqrt[m]{a}}{\sqrt[n]{b}}$, the denominator is said to be **rationalized** when we multiply both numerator and denominator by any quantity which renders the denominator rational.

Thus
$$\frac{\sqrt[3]{5}}{\sqrt{7}} = \frac{\sqrt{7}\sqrt[3]{5}}{7}.$$

Of course we might rationalize the numerator instead of the denominator, and cases will arise where this is the preferable transformation to make; but in general in division, if either dividend or divisor must be sacrificed as regards simplicity of form, it is advantageous to secure a simple divisor even at the expense of the dividend. The student will discover the reason for this if he will make a comparative study of the two cases; one in which the divisor is a simple quantity, and another in which the or is such a quantity as

$$3.14159, \text{ or } 1 + \sqrt{2}, \text{ or } x - \frac{x^2}{2} + \frac{x^3}{3} - \frac{x^4}{4} + \cdots.$$

57. The commonest case requiring rationalization of the denominator is that of a fraction in which the denominator is a binomial with one or both terms surds of the second order.

If we have $\dfrac{\sqrt[m]{a}}{\sqrt{b}+\sqrt{c}}$, the expression immediately becomes a surd with a rational denominator by multiplying both numerator and denominator by $\sqrt{b}-\sqrt{c}$.

Likewise, $\dfrac{\sqrt[m]{a}}{\sqrt{b}-\sqrt{c}}$ is rationalized by multiplying both numerator and denominator by $\sqrt{b}+\sqrt{c}$.

When two binomial quadratic surds differ only in the sign which connects the terms, they are said to be **conjugate**.

Thus $\sqrt{b}+\sqrt{c}$ and $\sqrt{b}-\sqrt{c}$ are conjugate. The product of two conjugate surds is a rational quantity; for $(\sqrt{b}+\sqrt{c})(\sqrt{b}-\sqrt{c}) = (\sqrt{b})^2 - (\sqrt{c})^2 = b - c$.

58. For the purposes of a subsequent article in the theory of equations, we shall now prove that the square root of a rational quantity cannot be partly rational and partly a quadratic surd.

Let us assume that the contrary is true, and suppose

$$\sqrt{a} = b + \sqrt{c};$$

squaring these equal expressions, we have

$$a = b^2 + 2b\sqrt{c} + c,$$

and therefore $\quad \sqrt{c} = \dfrac{a - c - b^2}{2b}.$

As the result of our assumption, we have the statement that a surd is equal to a rational quantity, which is impossible; therefore the assumption must be discarded.

ALGEBRA.

EXAMPLES.

1. Reduce the following expressions to their simplest forms:

$(x-y)[(x^2-y^2)(x-y)]^{\frac{1}{2}}$; $\quad \sqrt{338\,p^{\frac{2}{3}}q^3}$; $\quad \dfrac{m^{-1}}{n^0}\sqrt{63\,m^2 n^{-2}}$.

2. Write the simplest expression for the sum of
$$\sqrt{a^3},\ -\sqrt{ab^2},\ (a-b)\sqrt{a}.$$

3. Find the sum of $\dfrac{\sqrt{c+d}+\sqrt{c-d}}{-\sqrt{c-d}+\sqrt{c+d}}$ and its reciprocal.

4. Write the square of $\sqrt{\frac{1}{2}}+3\sqrt{\frac{1}{3}}$.

5. Divide $\sqrt[4]{\dfrac{x}{y}}$ by $\sqrt{\dfrac{y}{x}}$.

6. Prove that $\dfrac{\sqrt[m]{x}}{\sqrt[m]{y}}=\sqrt[m]{\dfrac{x}{y}}$, and state as a theorem.

7. Write the following fractions with their denominators rationalized:

$$\dfrac{\sqrt{5}-1}{7+\frac{1}{2}\sqrt{5}};\quad \dfrac{\sqrt{4+x^2}-2}{\sqrt{4+x^2}-3};\quad \dfrac{a^2}{\sqrt{a^2+b^2}+b}.$$

8. Write the quotient of $a-b \div \sqrt{a}-\sqrt{b}$.

9. The answer to a certain problem in dynamics was given as $10\sqrt{2}-\sqrt{3}$; a student obtained as the answer, $5\sqrt{2}(\sqrt{3}-1)$. Do the two expressions agree?

10. "If two expressions are equal, their like powers are equal." Establish the proposition.

SURDS.

11. State and prove a theorem corresponding to the one given in Ex. 10.

12. Why is not $\sqrt{a} + \sqrt{b} = \sqrt{a+b}$?

13. Given $a\sqrt[n]{b}$; it is required to derive an equivalent form in which the rational factor shall be under the radical sign. State a rule for all such cases.

14. Establish independently of the rule for multiplication,

(1) the rule for division of surds of the same order;
(2) the rule for division of surds of different orders.

CHAPTER VII.

IMAGINARY QUANTITIES.

59. In considering the subject of multiplication we found that the product of an *even* number of negative factors is a positive quantity. If the factors are all equal, we have merely a special case symbolically expressed by $(-a)^m$, in which a is positive and m is an even number.

Now since
$$(-a)^m = +a^m,$$
it follows that such a form as $-a^m$, a being positive and m being even, cannot arise by the ordinary process of raising a quantity to an even power. Consequently, the reverse operation, namely the extraction of an even root of a negative quantity, cannot be performed.

Nevertheless it is convenient to recognize such expressions as *indicated even roots* of negative quantities, and to call them **imaginary quantities** as distinguished from all other quantities, which are spoken of as **real**.

60. The commonest imaginary quantity is an indicated square root of a negative quantity.

By $\sqrt{-a}$, in which a itself is positive, we shall not mean any possible arithmetical operation, but shall define the expression by the statement,
$$(\sqrt{-a})(\sqrt{-a}) = -a.$$

IMAGINARY QUANTITIES.

61. Just as a surd admits of simplifying when it can be resolved into a rational factor and an irrational one, so the imaginary quantity $\sqrt{-a}$ may be expressed as the product of a real factor and an imaginary factor and thus simplified.

For by definition,
$$(\sqrt{-1})(\sqrt{-1}) = -1;$$
therefore $(\sqrt{a}\sqrt{-1})(\sqrt{a}\sqrt{-1}) = (-1)\sqrt{a}\sqrt{a};$

that is, $(\sqrt{a}\sqrt{-1})^2 = (-1)a = -a;$

hence $\sqrt{a}\sqrt{-1} = \sqrt{-a}.$

But this result is what we obtain if we treat $\sqrt{-a}$ as surds are treated, writing
$$\sqrt{-a} = \sqrt{a(-1)} = \sqrt{a}\sqrt{-1}.$$

The imaginary factor $\sqrt{-1}$ is called the **imaginary unit**.

An expression of the form $a + b\sqrt{-1}$, in which a and b are real quantities, is called a **complex number**. If the real term is zero, the expression is called a **pure imaginary**.

Since $\sqrt{-a} = \sqrt{a}\sqrt{-1}$, it will be seen that rules for the treatment of imaginaries follow the rules for surds, and need not be here repeated.

62. Two complex expressions differing only in the sign preceding the imaginary part are said to be **conjugate**.

We notice two important properties of conjugate imaginaries:

1. *The sum of two conjugate imaginaries is a real quantity;*

for $(a + b\sqrt{-1}) + (a - b\sqrt{-1}) = 2a.$

2. *The product of two conjugate imaginaries is a real quantity;*

for $(a + b\sqrt{-1})(a - b\sqrt{-1}) = a^2 - (b\sqrt{-1})(b\sqrt{-1})$
$= a^2 - b^2(-1) = a^2 + b^2.$

63. If the complex quantity $a + b\sqrt{-1}$ is equal to zero, a and b are individually equal to zero;
for, from the supposition

$$a + b\sqrt{-1} = 0,$$

we have $\quad\quad\quad\quad a = -b\sqrt{-1};$

and therefore $\quad\quad a^2 = b^2(\sqrt{-1})^2;$

that is, $\quad\quad\quad\quad a^2 = -b^2,$

and hence $\quad\quad a^2 + b^2 = 0.$

But a^2 and b^2 are each positive, and the sum of two positive quantities cannot be zero unless each of the quantities is zero.

64. The preceding article affords a fair specimen of the kind of reasoning which prevails in mathematics. (See last chapter.)

As regards form, the proposition follows the type form,

If a is β, γ is δ,

in which the condition, *if a is β*, is specialized as 'if the complex quantity $a + b\sqrt{-1}$ is equal to zero,' and the conclusion, *γ is δ*, is specialized as 'a and b are individually equal to zero.'

The process of establishing the conclusion, *γ is δ*, is merely a matter of setting down in algebraic shorthand a few propositions which are in accordance with the primary principles of quantitative relations.

A first conclusion is reached, and this in turn becomes a condition which compels a second conclusion. By repetitions of this process, the final conclusion is reached.

IMAGINARY QUANTITIES.

EXAMPLES.

1. For the following expressions write equivalent ones with real denominators:

$$\frac{x - y\sqrt{-1}}{\sqrt{x} + \sqrt{-y}}; \quad \frac{2 - \sqrt{-3}}{1 - \sqrt{-2}}; \quad \frac{a + \sqrt{a^2}}{a + \sqrt{-a^2}}.$$

2. Simplify the expressions:

$$2 + \sqrt{-63}; \quad \sqrt{x^{-2}} + \sqrt{-x^2}; \quad \left(\frac{1 + \sqrt{-1}}{\sqrt{2}}\right)^4.$$

3. Simplify $3\sqrt{-20} + 2\sqrt{-27} - \sqrt{-50}$.

4. Divide $\sqrt{18}$ by $\sqrt{-81}$.

5. Rationalize the denominator of

$$\frac{a + b\sqrt{-1}}{a - \sqrt{-b}} - \frac{a - \sqrt{-b}}{a + b\sqrt{-1}}.$$

CHAPTER VIII.

FACTORS.

65. When an algebraic expression and one of its factors are given, the rules for division enable us to find a third expression, which, multiplied by the given factor, produces the given expression.

In factoring proper, however, no factor is given to start with, and we must then discover factors either by inspection or by trial or by classifying the given expression as some recognizable form whose factors are known.

Certain kinds of factors are often more desirable than other kinds. In order to discuss them, we notice that algebraic expressions admit of a classification based on the absence from the denominator of letters with positive exponents; or, what comes to the same thing, on the absence from the numerator of letters with negative exponents.

If there is no letter symbol of quantity with a negative exponent in the numerator of any term of the given expression, the expression is said to be **integral**.

For example, $ax + b$ is an integral expression in each one of the letters a, x, b; $\dfrac{x}{a^2}$ is integral as regards x, but not as regards a; $\dfrac{a}{x^{-2}} + bx + c$ is an integral expression as regards x, because x^{-2} in the denominator may be written x^2 in the numerator.

As we have already seen, an expression is *rational* if

FACTORS.   57

none of its terms contain quantities with fractional exponents.

It follows that, in order to be both rational and integral, an expression written with all of its letters in the numerator must not contain negative or fractional exponents.

When a rational integral expression is given, we are to understand, if nothing is said to the contrary, that by its factors are meant rational integral factors. This limitation makes the work of factoring definite and often short; for it is obvious that if factors with negative or fractional exponents are admitted, there is no limit to the variety of factors into which an expression may be decomposed or resolved.

Thus the rational integral monomial $3\,ax^2y$ is seen to have the numerical factor 3, the rational integral factors a and y, together with two other rational integral factors x, x; but if the factors are not thus limited, y, for instance, may be further resolved into $y^{\frac{1}{2}}, y^{\frac{1}{2}}$; or $y^{\frac{1}{3}}, y^{\frac{2}{3}}$; or y^{-2}, y^3, etc.; and so with the others.

66. Among expressions which are to be recognized as factorable into particular binomial factors, which may or may not be rational and integral, are the following:

1. A trinomial so constituted that two of its terms have like signs and the third term is plus or minus twice the product of the square roots of the other two terms is factorable.

That is, $a + b \pm c$ is factorable if a and b have like signs, and if $c = 2\sqrt{ab}$.

We then have, if a and b are both positive,

$$a + b + 2\sqrt{ab} = (\sqrt{a} + \sqrt{b})(\sqrt{a} + \sqrt{b});$$
and $\quad a + b - 2\sqrt{ab} = (\sqrt{a} - \sqrt{b})(\sqrt{a} - \sqrt{b}).$

If a and b are both negative, we remove the factor -1 and proceed as before; then

$$-a-b-2\sqrt{ab} = -1(a+b+2\sqrt{ab})$$
$$= -1(\sqrt{a}+\sqrt{b})(\sqrt{a}+\sqrt{b});$$
also $-a-b+2\sqrt{ab} = -1(\sqrt{a}-\sqrt{b})(\sqrt{a}-\sqrt{b}).$

2. An expression consisting of the difference of two quantities is factorable.

That is, $a-b$ is factorable,

for $\qquad a-b = (\sqrt{a}+\sqrt{b})(\sqrt{a}-\sqrt{b}).$

3. An expression consisting of the sum of two quantities is factorable.

That is, $a+b$ is factorable,

for $\qquad a+b = (\sqrt{a}+\sqrt{b}\sqrt{-1})(\sqrt{a}-\sqrt{b}\sqrt{-1}).$

As examples of these three cases, $9a^2x^2 + 6axy + y^2$ has for its factors, $3ax+y$ and $3ax+y$, because the terms $9a^2x^2$ and y^2 have like signs and the third term $6axy$ is twice the product of the square roots of the other terms.

$3a^3x^4 - 2abx^2y\sqrt{3a} + b^2y^2 = (ax^2\sqrt{3a} - by)^2$ for the same reasons which were found in the first example.

$4x^2 - 3y^2$ has for its factors $2x + \sqrt{3}y$ and $2x - \sqrt{3}y$.

Similarly, $x^2 + 4y^2$ may be resolved into the factors $x + 2y\sqrt{-1}$ and $x - 2y\sqrt{-1}$.

The recognition of the factorability of an expression of the form of $ax^2 + by^2$ is of importance in analytic geometry.

67. It is to be noticed in the three cases just presented that the binomial factors are rational and integral as regards certain letters only when those letters are of *even dimensions* in the given expression.

Thus we have in general,

$$ax^m + by^n \pm 2\sqrt{ab}x^{\frac{m}{2}}y^{\frac{n}{2}} = \left(\sqrt{a}x^{\frac{m}{2}} \pm \sqrt{b}y^{\frac{n}{2}}\right)^2.$$

Now $\sqrt{a}x^{\frac{m}{2}} \pm \sqrt{b}y^{\frac{n}{2}}$ is a rational integral expression in x and y if both m and n are even numbers; but if either m or n is an odd number, the expression is not integral.

It should be noticed that the expression is rational and integral in x and y; a and b are viewed as coefficients and hence there is no objection to the surds $\sqrt{a}$ and $\sqrt{b}$ appearing in the factors.

Again,
$$ax^m - by^n = \left(\sqrt{a}x^{\frac{m}{2}} + \sqrt{b}y^{\frac{n}{2}}\right)\left(\sqrt{a}x^{\frac{m}{2}} - \sqrt{b}y^{\frac{n}{2}}\right),$$
and
$$ax^m + by^n = \left(\sqrt{a}x^{\frac{m}{2}} + \sqrt{-1}\sqrt{b}y^{\frac{n}{2}}\right)\left(\sqrt{a}x^{\frac{m}{2}} - \sqrt{-1}\sqrt{b}y^{\frac{n}{2}}\right),$$

and these factors are integral expressions in x and y if m and n are even numbers; but if either m or n is an odd number, the factors are not integral.

68. Other forms which may be immediately factored are: $x^m - y^m$, m being *any* integer; $x^m + y^m$, m being an odd number.

One factor in each case is a linear homogeneous binomial; it is left to the student to discover what the binomial is, and to note the characteristics of the other factor as regards degree, homogeneity, symmetry, factorability, etc.

69. The general non-homogeneous quadratic expression in one quantity, as x, requires especial attention

because of its very frequent occurrence in elementary mathematics. The expression in question may be written $ax^2 + bx + c$; evidently it is not the square of an integral linear expression unless the coefficients a, b, c, are so related that
$$b = 2\sqrt{ac},$$
and therefore
$$c = \frac{b^2}{4a}.$$

Now if we are at liberty to change the expression by adding $\frac{b^2}{4a} - c$ to it, we have
$$ax^2 + bx + c + \left(\frac{b^2}{4a} - c\right) = ax^2 + bx + \frac{b^2}{4a}$$
$$= \left(x\sqrt{a} + \frac{b}{2\sqrt{a}}\right)^2.$$
Similarly,
$$ax^2 - bx + c + \left(\frac{b^2}{4a} - c\right) = \left(x\sqrt{a} - \frac{b}{2\sqrt{a}}\right)^2.$$

The quadratic expression $ax^2 + bx + c$ sometimes admits of resolution into two unequal linear factors, no change being made in the absolute term. Suppose the factors are $mx + p$ and $nx + q$; then
$$ax^2 + bx + c = mnx^2 + (qm + pn)x + pq,$$
and
$$mn = a, \quad qm + pn = b, \quad pq = c.$$

In the case of the simpler trinomials, which admit of resolution into two unequal linear factors, inspection will usually show the values of m, n, p, q.

For example, $x^2 + 5x + 6 = (x + 3)(x + 2)$.

mn is unity in this case, and p and q must be numbers such that their product shall be six and their sum five.

FACTORS. 61

Again, $x^2 + 2x - 3 = (x - 1)(x + 3)$.

mn is unity in this example also; and either p or q must be a negative quantity, since $pq = -3$; and since the sum of p and q is two, the positive quantity must be three, and the negative quantity must be unity with the minus sign.

70. H. C. F. The **highest common factor** of two or more expressions is the factor of highest degree which will exactly divide each of them. On the one hand, it cannot contain any factor not found in each of the expressions; and on the other hand, it must contain every factor common to each of the expressions, and contain it as many times as it is found in that expression which contains it the fewest number of times.

If the given expressions are analyzed into their factors, the H. C. F. may be at once formed. If the given expressions are two polynomials, the H. C. F. is usually found without preliminary factoring.

Let N and n represent the two polynomials having no common monomial factors, and suppose N of higher degree than n.

If n will exactly divide N, it is of course the required H. C. F.; but, in general, one of the polynomial expressions will not exactly divide the other.

Now let the operation of dividing N by n be begun. The first remainder will consist of the dividend N minus some multiple of the divisor n; call the multiple qn. Then this remainder $N - qn$ will be exactly divisible by the H. C. F., because each term of it is divisible by the H. C. F.

Let the division of N by n be continued until the remainder is of a lower degree than the divisor. The

62　　　　　ALGEBRA.

H. C. F. must exactly divide this last remainder for the same reason that it divides the first remainder. Therefore, we may as well begin anew, and divide the divisor n by this remainder, precisely as if they were the two expressions given at the outset.

By continuing this operation of dividing the last divisor used by the last remainder found, we come at length upon a remainder and divisor such that the remainder will exactly divide the divisor used as a dividend. In this case the last remainder, used as a divisor, is the H. C. F. of the given expressions.

The conclusion just stated is based on the supposition that N and n have an H C. F. higher than the zero degree. However, if the process of divisions has to be continued until the last remainder used as a divisor is a linear expression and is not exactly contained in the dividend, we must conclude that N and n have no H. C. F.

To illustrate the process above described, let it be required to find the H. C. F. of

$$x^3 + 4x^2 + 6x + 4 \text{ and } x^2 + 3x + 2.$$

In the first place, we observe that these two expressions have no common monomial factor. If they had such a factor, we should take it out and set it aside as one of the factors of the H. C. F.

In the second place, since the H. C. F. will contain only factors common to the two expressions, we may introduce any desired factor into one of the expressions. Since it is not introduced into the other, it is not a factor of the H. C. F., and hence it will not affect the result. In this particular example, however, we shall have no occasion to introduce any factor at any stage of the operation.

FACTORS.

Performing the divisions, we have the following operation:

$$
\begin{array}{r}
x^3 + 4x^2 + 6x + 4\,\underline{(x^2 + 3x + 2} \\
\underline{x^3 + 3x^2 + 2x}\;\;(x+1 \\
x^2 + 4x + 4 \\
x^2 + 3x + 2
\end{array}
$$

$$
\begin{array}{r}
x^2 + 3x + 2\,\underline{(x+2} \\
\underline{x^2 + 2x}\;\;(x+1 \\
x + 2 \\
x + 2
\end{array}
$$

We have thus found that the required H. C. F. is the divisor $x + 2$.

The student should compare this operation step by step with the theory as outlined.

EXAMPLES.

1. Factor the following expressions:

$9x^2 - 25y^2$; $x^3 + y^3$; $4y^2 + 16x^2$.
$3 + a^3$; $x^{2m} + y^{2m} - 2x^m y^m$; $p^{10} - q^5$.

2. Given $(ax^2 + bxy + cy^2)$. Place the factor x^2 outside of the parentheses. What will be the degree of the factor remaining within the parentheses? Will it be homogeneous?

3. Write the homogeneous expression of the nth degree in x and y,

$$p_1 x^n + p_2 x^{n-1} y + p_3 x^{n-2} y^2 + \cdots + p_{n-1} x^2 y^{n-2} + p_n x y^{n-1} + p_{n+1} y^n,$$

as an expression of the nth degree in $\dfrac{y}{x}$.

Note. Thus far we have used only letters of the English alphabet to represent quantity; but it will often be convenient to use Greek letters, and also English letters with subscripts and primes.

Thus a' is read, 'a prime'; a'' is read, 'a second,' or 'a double prime'; p_0 is read, 'p sub zero'; n_3 is read, 'n sub three.' The student will readily see the advantage gained by the use of the coefficient symbols p_1, p_2, $\cdots$ p_{n+1}, instead of such coefficients as a, b, c, $\cdots$.

4. Employ the method of Art. 70 to determine whether $x^3 - 2x^2 - 6x + 4$ and $3x^2 - 4x - 6$ have an H. C. F.

5. Find the H. C. F. of $x^3 - 3x^2 + 2x - 6$ and $3x^2 - 6x + 2$.

CHAPTER IX.

EQUATIONS.

71. Thus far we have discussed transformations of algebraic expressions, the transformations being effected according to primary laws and by means of an adopted symbolism.

We have now to consider simultaneous transformations of two algebraic expressions which are equal to each other.

The statement that two expressions are equal to each other is made by means of the verb symbol $=$ ('*equals*'). The statement itself is called an **equation**.

Suppose that A represents any algebraic expression, and A' any other algebraic expression. The statement $A = A'$ means that the two expressions, although different in *form*, are the same in *value*.

The expressions A, A' are called the **members** or **sides** of the equation.

72. Law of the equation. The fundamental law governing transformations of equations is as follows:

If the same operation be performed on two equal expressions, the resulting expressions will be equal.

The law itself rests on the still more fundamental fact of experience, that magnitudes or values or quantities are independent of the method of measuring them and of the method of expressing their measure.

73. Just as various adjectives were found convenient in describing and classifying various kinds of expressions, so we shall use the same adjectives to describe the equations containing the expressions to which the various adjectives are applied.

To illustrate this new use of old terms, and also to introduce certain new terms, let us consider the statement
$$ax = b. \tag{1}$$

According to the law given in Art. 72, if we subtract the same any quantity from each member of equation (1), the expression for the remainders will be equal, and hence will constitute an equation. Subtracting b from both members, we have
$$ax - b = 0. \tag{2}$$

The advantage of subtracting b in this case is that the expression for the right-hand remainder is zero; and we thus have the expression for the left-hand remainder equal to zero. In general, if $A = A'$ is the initial form of any equation, we shall bring it to the form $A - A' = 0$ before beginning its study.

It should be noted that the operation which is actually performed is that of subtracting the second member from each member of the equation; but it is often described as **transposing** A' to the left-hand side of the equation.

Besides writing equation (1) in the form numbered (2), the law of the equation permits us to divide each member of equation (1) by the same any quantity. Dividing by a, we have
$$x = \frac{b}{a}; \tag{3}$$

so that, observing the conditional form,

if $\qquad a$ is β, γ is δ,

we have,

if $\qquad ax - b = 0,\ x = \dfrac{b}{a}.$

74. Root of an equation. If at the outset we did not know the value of x, and if we imagined that it might have a variety of values, we have found that it has one and only one value, and we have obtained this value as an expression in terms of the other symbols of quantity a and b, each of which is supposed to be known.

In arriving at the conclusion $x = \dfrac{b}{a}$, we **solve** the equation for the unknown quantity x. The expression $\dfrac{b}{a}$ is called the **root** of the equation.

We define a root of an equation as a value of the unknown quantity, which, being substituted for the unknown quantity in the given equation, will make the equation take such a form as to state that an expression is equal to itself. This is called **satisfying** the equation. Thus, if we replace x by $\dfrac{b}{a}$ in equation (1), we have

$$a\left(\dfrac{b}{a}\right) = b;$$

$$\therefore b = b.$$

The reason why we define a root as *a* value, instead of *the* value of the unknown quantity, will be seen later.

The term '*root*' as here used must not be confused with the term when used for one of the equal factors of a quantity.

75. Linear equations. The expression $ax - b$, in the equation $ax - b = 0$, is seen to be a linear non-homogeneous expression in x; the equation is therefore called a linear non-homogeneous equation in x. It is also called a **literal** or **general** equation because the general quantities a and b enter into it. If a and b receive particular numerical values, the equation is then said to be **numerical**.

Thus the equations $x + 7 = 0$, $3x - \tfrac{1}{5} = 0$, are numerical equations.

Finally, we notice that the solution of the general equation of the first degree in one unknown quantity, by virtue of its being general, includes the solution of any particular (numerical) equation belonging to the type equation $ax - b = 0$; and if we translate the statement $x = \dfrac{b}{a}$, we have a proposition describing the root. The formal translation is: If an equation of the first degree in one unknown quantity be written in the form $ax + b = 0$, the root of the equation is the quotient obtained by dividing the absolute term with its sign changed by the coefficient of the unknown quantity.

According to this theorem, the root of the equation $3x - \tfrac{1}{5} = 0$ is $\tfrac{1}{15}$; the root of the equation $x + 7 = 0$ is -7.

76. In the preceding article we have been careful to describe the equation $ax - b = 0$ as an equation in *one* unknown quantity. Suppose this limitation removed, so that we have one equation of the first degree in two (or more) unknown quantities; for example, let the equation be

$$ax + by + c = 0, \qquad (4)$$

in which x and y are unknown.

For the moment, let us regard (4) as an equation in x alone. Solving for x as in the previous article, we have

$$x = \frac{-(by+c)}{a}. \tag{5}$$

Now an unknown quantity cannot be said to be known when it is expressed in terms of another unknown; and since this value of x contains the unknown y, it is plain that if we have no means of determining y, if we grant that it may have this or that or any value, we then cannot determine x; it also has various values, changing when y changes.

It thus appears that a single equation in two unknown quantities does not admit of solution in the same sense that a single equation in one unknown quantity admits of it.

77. Suppose now that, besides having equation (4) concerning x and y, we have the accompanying statement,

$$a'x + b'y + c' = 0. \tag{6}$$

From (4), as already noticed,

$$x = \frac{-(by+c)}{a};$$

similarly, from (6),

$$x = \frac{-(b'y+c')}{a'}. \tag{7}$$

Since (4) and (6) relate to the same quantity, represented by x, expressions (5) and (7) for this quantity must be equal to each other. That is,

$$\frac{-(by+c)}{a} = \frac{-(b'y+c')}{a'}. \tag{8}$$

Equation (8) is now an equation in the one unknown quantity y. Several operations performed mentally and justified by the law of the equation give us

$$y = \frac{\dfrac{c'}{a'} - \dfrac{c}{a}}{\dfrac{b}{a} - \dfrac{b'}{a'}} = \frac{ac' - ca'}{a'b - ab'}, \qquad (9)$$

and as this y is the y of (4), we may return to that equation and substitute expression (9) for y; we then have

$$ax + b\left(\frac{ac' - ca'}{a'b - ab'}\right) + c = 0;$$

hence
$$x = \frac{-\left[c + b\left(\dfrac{ac' - ca'}{a'b - ab'}\right)\right]}{a}.$$

From this it appears that with two linear equations in two unknown quantities we are able to find the values of the unknown quantities in terms of known coefficients.

78. When associated equations are statements about the same quantity or quantities, they are said to be **simultaneous**.

Thus equations (4) and (6) in the preceding article are simultaneous. If the x and y of (6) had not been the x and y of (4), we could not have made the combinations which were made.

Again, when associated equations are different statements about the same quantity or quantities, they are said to be **independent**. By different statements we do not mean that they are contradictory, but that one cannot be derived from the other. If a', b', c' in (6) are so related to a, b, c in (4) that $a' = na$, $b' = nb$, $c' = nc$, the

EQUATIONS.

two equations are not independent; for, by dividing both members of (6) by n, we should have (4).

79. Summarizing results thus far reached, one equation in one unknown quantity is sufficient for the determination of that unknown; and two equations in two unknown quantities are sufficient for the determination of the two unknown quantities; but one equation in two unknown quantities is not sufficient for the determination of those quantities.

In general, if we have n unknowns, we need n simultaneous independent equations in order to find the values of the unknowns.

Suppose the equations are:

$$a_1x + b_1y + c_1z + \cdots + q_1 = 0 \qquad (1)$$
$$a_2x + b_2y + c_2z + \cdots + q_2 = 0 \qquad (2)$$
$$\cdot \quad \cdot \quad \cdot \quad \cdot \quad \cdot \quad \cdot$$
$$a_nx + b_ny + c_nz + \cdots + q_n = 0 \qquad (n)$$

in $x, y, z, \cdots$ to n unknown quantities.

From equation (1) we may write the value of x in terms of the coefficients of the equation and the other unknown quantities. Substituting this value of x in the $\overline{n-1}$ remaining equations, these equations will contain only $y, z, \cdots$ to $\overline{n-1}$ unknown quantities.

From (2) we may now write the value of y, and substitute it in the $\overline{n-2}$ remaining equations, which will then contain only

$$z, \cdots \text{to } \overline{n-2} \text{ unknown quantities.}$$

Proceeding in this way, we arrive at length at the nth equation, which will contain only one unknown quantity.

If we had started with $\overline{n-1}$ (or fewer) equations in

n unknown quantities, we should reach, by the process above described, a single equation in two (or more) unknown quantities; and, as we have seen, such an equation cannot be solved in the sense that we find one root or a few roots.

It is to be noticed that the order of using the n equations, and likewise the order of eliminating the unknown quantities, is of no consequence.

Thus we might have begun with the nth or $\overline{n-2}$th equation, and solved first for y or z.

80. In the above argument the equations used are all linear, because we have thus far discussed only linear equations; but if they were of different degrees, and we knew how to solve equations of higher degrees than the first, the outcome would be the same: we should need as many equations as we have unknown quantities, in order to find the values of the unknown quantities.

81. The inquiring student will now naturally ask, what would happen if we had *more* equations than we have unknown quantities?

To answer this question, let us consider the simplest possible case.

Suppose we have two simultaneous equations in one unknown quantity, and let them be

$$ax + b = 0 \text{ and } a'x + b' = 0.$$

From the first $\quad x = -\dfrac{b}{a},$

and from the second $\quad x = -\dfrac{b'}{a'};$

hence $\quad \dfrac{b}{a} = \dfrac{b'}{a'};$

and if $\qquad b' = nb,$

we have also $\qquad a' = na;$

and, consequently, the equations are not independent.

But if it be insisted upon that they are independent, we then must not equate $\dfrac{b}{a}$ and $\dfrac{b'}{a'}$; we can only say that the results are discordant and that by some means or other, with the two values of x, $\dfrac{b}{a}$ and $\dfrac{b'}{a'}$ as material, we must determine the *most probable* value of x. The theory of the determination of the most probable values of n unknown quantities, when more than n equations are given, belongs to a branch of mathematics far in advance of elementary algebra and can only be referred to here. It may be added that the problem of the solution of n equations involving fewer than n unknown quantities, is a problem of constant occurrence in certain branches of science.

82. We pass now to the study of the equation which arises when a quadratic expression in any one quantity is equated to zero.

The most general form of such an equation is

$$ax^2 + bx + c = 0, \qquad (1)$$

in which the coefficients a, b, c represent any quantities positive or negative, integral or fractional, rational or irrational.

In attempting to factor the expression $ax^2 + bx + c$ (Art. 69), we observed that if we are permitted to add $\dfrac{b^2}{4a} - c$ to the expression, it becomes a perfect square.

The law of the equation does permit us to add any quantity we please to both members of the equation. Adding $\frac{b^2}{4a} - c$ to each member of equation (1), we have

$$ax^2 + bx + c + \frac{b^2}{4a} - c = \frac{b^2}{4a} - c; \qquad (2)$$

that is,
$$\left(\sqrt{a}x + \frac{b}{2\sqrt{a}}\right)^2 = \frac{b^2}{4a} - c,$$

and hence
$$\sqrt{a}x + \frac{b}{2\sqrt{a}} = \pm\sqrt{\frac{b^2}{4a} - c};$$

hence
$$x = -\frac{b}{2a} \pm \frac{1}{2a}\sqrt{b^2 - 4ac}. \qquad (3)$$

Since we have now solved for x in the most general equation of the second degree in one unknown quantity, it becomes important to make a careful study of the result.

However, instead of using the form just obtained, there will be found to be a certain advantage in first dividing both members of equation (1) by the coefficient of x^2. We then have

$$x^2 + \frac{b}{a}x + \frac{c}{a} = 0. \qquad (4)$$

For convenience, let $\frac{b}{a} = p$ and $\frac{c}{a} = q$.

Equation (4) then becomes

$$x^2 + px + q = 0, \qquad (5)$$

an equation just as general as equation (1), but with unity for the coefficient of the term of the highest degree in x.

To render the first member of equation (5) a perfect square, we need to add $\frac{p^2}{4} - q$ to it.

Then $\quad x^2 + px + q + \frac{p^2}{4} - q = \frac{p^2}{4} - q;$ \hfill (6)

that is, $\quad\quad\quad \left(x + \frac{p}{2}\right)^2 = \frac{p^2}{4} - q,$

and hence $\quad\quad x + \frac{p}{2} = \pm \sqrt{\frac{p^2}{4} - q};$

therefore $\quad\quad\quad x = -\frac{p}{2} \pm \sqrt{\frac{p^2}{4} - q}.$ \hfill (7)

We notice that

(1) there are *two* roots, as contrasted with the one root of the corresponding first degree equation; in reading expression (7), the double sign must be read, 'plus *and* minus,' not 'plus *or* minus';

(2) the roots are binomial in form, and differ only in the sign preceding the surd term;

(3) the first term of each root is half the coefficient of the first power of x, with its sign changed; and the second term is the square root of the square of half the coefficient of the first power of x, minus the absolute term.

In speaking of the absolute term and the coefficient of the first power of x, we mean each of these quantities taken with the sign which precedes it.

83. Sum and product of the two roots. Denoting the two roots of equation (5) by α and β, and writing expression (7) as two separate expressions, we have

$$\alpha = -\frac{p}{2} + \sqrt{\frac{p^2}{4} - q},$$

$$\beta = -\frac{p}{2} - \sqrt{\frac{p^2}{4} - q}.$$

Now if we add these expressions, we have

$$\alpha + \beta = -p;$$

that is, *the sum of the roots is the coefficient of the first power of x, with its sign changed.*

Again, if we multiply the roots together,

$$\alpha\beta = \left(-\frac{p}{2} + \sqrt{\frac{p^2}{4} - q}\right)\left(-\frac{p}{2} - \sqrt{\frac{p^2}{4} - q}\right);$$

or, observing that we have the product of the sum and difference of two quantities, and performing the indicated operation,

$$\alpha\beta = q;$$

that is, *the product of the roots is the absolute term.*

The student should bear it in mind that the two theorems just obtained assume that the coefficient of x^2 is positive and is unity; that is, that the equation is brought to the form

$$x^2 + px + q = 0$$

before solving. If the equation were in the form

$$-mx^2 + nx = r,$$

it would evidently not be true that the sum of the roots is the coefficient of the first power of x, with its sign changed, that is, $-n$. Neither would the product of the roots equal r, the absolute term.

84. Since p and q in equation (7) represent any quantities whatever, it is obvious that the quantity beneath the radical sign will be sometimes positive, sometimes negative, and sometimes it will be signless because its value is zero.

To get the exact relation of p and q for these three cases, we write expression (7) in the form

$$x = -\frac{p}{2} \pm \frac{1}{2}\sqrt{p^2 - 4q}. \qquad (8)$$

(1) If $p^2 > 4q$, the expression $p^2 - 4q$ is positive, and therefore $\sqrt{p^2 - 4q}$ is a real quantity, rational or irrational.

(2) If $p^2 < 4q$, $p^2 - 4q$ is negative, and therefore $\sqrt{p^2 - 4q}$ is an imaginary quantity.

(3) If $p^2 = 4q$, the second term of each root is zero; hence each root is in this case equal to $-\dfrac{p}{2}$.

From the foregoing it follows that

(1) either both roots are real or both imaginary;

(2) the condition that the roots shall be real is that
$$p^2 > \text{ or } = 4q;$$

(3) the condition that the roots shall be equal is that $p^2 = 4q$.

85. Further observation of expression (7) shows that if the absolute term q is zero, one of the roots is zero; namely, the root α, for then

$$\alpha = -\frac{p}{2} + \sqrt{\frac{p^2}{4}} = 0;$$

also, if q is zero, the other root becomes $-p$, for then

$$\beta = -\frac{p}{2} - \sqrt{\frac{p^2}{4}} = -p.$$

If p, the coefficient of the first power of x, is zero, the expressions for the roots become

$$\alpha = +\sqrt{-q},$$
$$\beta = -\sqrt{-q};$$

that is, the two roots are numerically equal, with opposite signs.

In case p is zero, equation (5) becomes

$$x^2 + q = 0.$$

This form is called an **incomplete** or **pure** quadratic; it may at once be written

$$x^2 = -q;$$

whence $\qquad x = \pm\sqrt{-q},$

as already seen.

Since $\qquad \alpha + \beta = -p,$
and $\qquad \alpha\beta = q,$

we may write equation (7)

$$x^2 - (\alpha + \beta)x + \alpha\beta = 0;$$

but the first member of this equation can be factored into

$$(x - \alpha)(x - \beta);$$

therefore we have

$$(x - \alpha)(x - \beta) = 0.$$

From this it follows that if a quadratic expression of the form $x^2 + px + q$ be resolved into two linear non-homogeneous factors of the form $x - \alpha$ and $x - \beta$, α and β are the roots of the equation obtained by equating the given quadratic expression to zero.

Conversely, if two roots are given and it is required to find the equation whose roots they are, we have only to subtract the given roots in succession from the unknown quantity, to indicate the product of the binomials thus formed, and to equate this product to zero.

Formula (7), the result of solving equation (5), should be committed to memory so that the student may write down at once the roots of any given quadratic equation. This solution of the general equation renders it unnecessary to go through the operation of solving any particular quadratic.

86. Any equation of the form
$$x^{2n} + px^n + q = 0$$
can evidently be solved as a quadratic; for if we put $z = x^n$, the given incomplete equation of the 2nth degree becomes
$$z^2 + pz + q = 0,$$
whence
$$z = -\frac{p}{2} \pm \sqrt{\frac{p^2}{4} - q},$$
and therefore
$$x = \sqrt[n]{-\frac{p}{2} \pm \sqrt{\frac{p^2}{4} - q}}.$$

87. Homogeneous equations. — A homogeneous equation is a homogeneous expression equated to zero.

Thus, $$ax^2 + bxy + cy^2 = 0 \qquad (9)$$
is the general homogeneous equation of the second degree in two unknown quantities.

If we divide equation (9) by x^2, we have
$$a + \frac{by}{x} + \frac{cy^2}{x^2} = 0,$$

and if we put $z = \dfrac{y}{x}$, this becomes

$$cz^2 + bz + a = 0,$$

which may now be solved as a quadratic in z; but it must be noticed that in thus finding the value of z we have merely found the ratio $\dfrac{y}{x}$; we do not know the individual values of x and y.

Some elementary algebra books give as examples of homogeneous equations such forms as

$$ax^2 + bxy + cy^2 = d\,;$$

that is, equations consisting of a homogeneous expression equated to an absolute term which is not zero; but it is to be observed that such an equation does not admit of the reduction and solution above given. The importance of having all the terms of the same dimensions will be appreciated by the student upon taking up the study of higher algebra and analytic geometry Consult Burnside and Panton's *Theory of Equations* (3d ed.), Art. 135; Loney's *Coördinate Geometry*, Art. 120; C. Smith's *Solid Geometry*, Art. 69.

88. The equation acquires its importance from the fact that when a problem is stated in the language of algebra, the statements are equations, and the solution of the problem requires combinations and transformations of equations. Numerous illustrative examples may be found in almost any elementary algebra text-book; it is not in the plan of the present work to give such examples, but rather to develop the theory of the quadratic equation, together with the more important principles relating to equations of higher degrees.

CHAPTER X.

RATIO.

89. Ratio is the relative magnitude of the measures of two quantities of the same kind.

This definition implies: (1) that we have two quantities of the same kind, as a handful of roses and another handful of roses, or a line AB and another line CD, or dynes and other dynes; (2) that a chosen unit of measure is applied to the quantities of the same kind: thus the line AB may contain the unit of measure a times while the line CD contains it b times; (3) that these expressions for the measures of the two quantities are then compared. The method of making this quantitative comparison is not by subtracting one expression from the other and then saying that one is so much more than the other; it is rather by dividing one expression by the other and then saying that one is so many times the other.

It is evident therefore that a ratio is always abstract. It takes the form of a fraction and is subject to the same rules to which fractions are subject; but whilst all ratios imply division as do all fractions, it by no means follows that all fractions are of the nature of ratios, for some fractions, *i.e* indicated divisions, are such that one term may be a concrete quantity.

The ratio expression $\frac{a}{b}$ is often written $a:b$ in order to distinguish it from an ordinary fraction. a is called the **antecedent** and b the **consequent**; the two together are called the **terms** of the ratio.

Whether $\frac{a}{b}$ can be expressed in integral form or not, it is important to realize that $\frac{a}{b}$ is to be regarded as a single expression representing the *relative* magnitude of two quantities; with the *absolute* magnitude of either quantity or term we are not concerned. In Part II. of this work the student will deal with ratios whose individual terms are infinitely small; but the relative magnitudes of these terms, that is the ratios, will be expressed by quantities that are in general finite.

90. Proportion. — If two ratios are equal to each other, the equation thus formed is called a **proportion**. If the two equal ratios are $\frac{a}{b}$ and $\frac{c}{d}$, the proportion is, by definition,

$$\frac{a}{b} = \frac{c}{d}; \qquad (1)$$

but to bring out the fact that it is an equality of ratios rather than of ordinary fractions, it is often written

$$a : b = c : d, \qquad (2)$$

and is read: 'the ratio of a to b equals the ratio of c to d'; or more briefly: 'a is to b as c is to d.'

The first and fourth terms of a proportion are called the **extremes**; the second and third terms are called the **means**.

91. In order to reach various theorems regarding a proportion, we only need to use form (1), performing certain simple operations in accordance with the law of the equation, and interpreting the results.

If each member of equation (1) be multiplied by bd, we have
$$\frac{abd}{b} = \frac{cbd}{d};$$
that is, $\qquad ad = bc.$

But ad is the product of the extremes of proportion (2), and bc is the product of the means. Hence, *the product of the extremes is equal to the product of the means.*

Conversely, if $\qquad ad = bc,$
the quantities a, b, c, d will be in proportion; for dividing each member of the equation of the condition by bd,
$$\frac{ad}{bd} = \frac{bc}{bd};$$
that is, $\qquad \dfrac{a}{b} = \dfrac{c}{d},$

and therefore $\qquad a : b = c : d.$

By similar steps, four additional conclusions are reached. These associated operations may be exhibited as follows:

if $\quad ad = bc,$ $\begin{cases} \dfrac{ad}{bd} = \dfrac{bc}{bd}, \\ \qquad \therefore a:b = c:d; \qquad (1) \\ \dfrac{ad}{cd} = \dfrac{bc}{cd}, \\ \qquad \therefore a:c = b:d; \qquad (2) \\ \dfrac{bd}{ad} = \dfrac{bd}{bc}, \\ \qquad \therefore b:a = d:c; \qquad (3) \\ \dfrac{ad}{ab} = \dfrac{bc}{ab}, \\ \qquad \therefore d:b = c:a; \qquad (4) \\ \dfrac{ab}{ad} = \dfrac{ab}{bc}, \\ \qquad \therefore b:d = a:c. \qquad (5) \end{cases}$

It is to be observed that any one of these conclusions may be taken as condition, and all the others may be concluded from it.

92. Mean proportional. — If the consequent of the first ratio of a proportion is equal to the antecedent of the second ratio, so that the proportion assumes the form

$$a : b = b : c, \qquad (3)$$

the common mean term is said to be a **mean proportional** between the other two terms; the third term is called a **third proportional** to the first two terms.

From proportion (3) we have

$$b^2 = ac,$$

whence $\qquad b = \pm \sqrt{ac}.$

The translation of this result affords another form of definition of a mean proportional.

93. Composition and division. — If proportion (2) be written in form (1) and unity be added to each member, we have

$$\frac{a}{b} + 1 = \frac{c}{d} + 1;$$

that is, $\qquad \dfrac{a+b}{b} = \dfrac{c+d}{d},$

and therefore $\quad a + b : b = c + d : d. \qquad (4)$

Also, subtracting unity from each member of (1),

$$\frac{a-b}{b} = \frac{c-d}{d};$$

therefore $\qquad a - b : b = c - d : d. \qquad (5)$

Proportion (4) is described as one in which the terms a, b, c, d are taken by **composition**; in (5) they are said to be taken by **division**.

Dividing each member of the equation

$$\frac{a+b}{b} = \frac{c+d}{d}$$

by the corresponding member of the equation

$$\frac{a-b}{b} = \frac{c-d}{d},$$

the result is

$$\frac{a+b}{a-b} = \frac{c+d}{c-d};$$

that is, $\qquad a+b : a-b = c+d : c-d.$ \qquad (6)

In proportion (6) the terms are said to be taken by composition and division

94. If several ratios are equal, we may find an important expression for any one of them in the following manner:

Suppose we have three equal ratios, represented by $\frac{a}{b}, \frac{c}{d}, \frac{g}{h}.$

For convenience, let each one of them equal some quantity as k.

From the equalities

$$\frac{a}{b} = \frac{c}{d} = \frac{g}{h} = k,$$

we obtain $\qquad a = bk, \quad c = dk, \quad g = hk.$

Then $\qquad a^m = b^m k^m, \quad c^m = d^m k^m, \quad g^m = h^m k^m,$

m being any exponent.

Adding the corresponding members of these equations,
$$a^m + c^m + g^m = k^m(b^m + d^m + h^m),$$

and therefore
$$k = \frac{(a^m + c^m + g^m)^{\frac{1}{m}}}{(b^m + d^m + h^m)^{\frac{1}{m}}};$$

but k represents each one of the ratios; hence

$$\frac{(a^m + c^m + g^m)^{\frac{1}{m}}}{(b^m + d^m + h^n)^{\frac{1}{m}}} = \begin{cases} \dfrac{a}{b}; \\ \dfrac{c}{d}; \\ \dfrac{g}{h}. \end{cases}$$

As a special case m may equal unity, and it follows that *if several ratios are equal, each one of them is equal to the ratio of the sum of the antecedents to the sum of the consequents.*

It is left to the student to state the corresponding theorems when $m = 2$, when $m = 3$, etc.

EXAMPLES.

1. If $a : b = c : d,$

show that $\dfrac{\sqrt{a^2 + b^2}}{b} = \dfrac{\sqrt{c^2 + d^2}}{d}.$

2. If $a : b = c : d,$

is $a^m : b^m = c^n : d^n$?

3. If $a : b = b : c,$

show that $a : a + b = a - b : a - c.$

4. Find a mean proportional between the quantities $(m+n)$ and $(m-n)$; also between the quantities $(m+n)^2$ and $(m-n)^2$.

5. Given $\dfrac{l}{a} = \dfrac{m}{b} = \dfrac{n}{c}$, with the relation $l^2 + m^2 + n^2 = 1$; show that $l = \dfrac{a}{\sqrt{a^2 + b^2 + c^2}}.$

CHAPTER XI.

PROGRESSIONS.

95. By a **series** is meant the algebraic sum of a number of expressions formed according to some common law. Each expression is called a **term**.

96. Arithmetic progression. A series is said to be an arithmetic series or **arithmetic progression** when the terms increase or decrease by a **common difference.**

Thus, $1 + 4 + 7 + 10 + \cdots$ is an arithmetic series in which each term is obtained from the preceding one by adding 3 to it; $7+5+3+1-1-3-\cdots$ is an arithmetic series in which the common difference is -2.

97. nth term. If we consider a general form for an arithmetic series, say the form
$$a + (a+d) + (a+2d) + (a+3d) + \cdots,$$
we notice that the coefficient of d in the third term is 2, of d in the fourth term is 3, and so on; that is, the coefficient of the common difference as it occurs in any term is one less than the number of that term.

Hence the general or nth term must be $a + (n-1)d$.

98. Sum of n terms. If s be the sum of n terms,
$$s = a + [a+d] + [a+2d] + \cdots$$
$$+ [a+(n-2)d] + [a+(n-1)d].$$

Writing these n terms in the reverse order,
$$s = [a+(n-1)d] + [a+(n-2)d] + \cdots \\ + [a+2d] + [a+d] + a.$$

Adding the first term and the nth term, the second term and the $\overline{n-1}$th term, and so on,
$$2s = [2a+(n-1)d] + [2a+(n-1)d] + \cdots \\ + [2a+(n-1)d] + [2a+(n-1)d] \\ = n[2a+(n-1)d].$$
$$\therefore s = \frac{n}{2}[2a+(n-1)d].$$

99. Arithmetic mean. When three quantities fulfil the requirements of an arithmetic progression, the middle one is said to be the **arithmetic mean** of the other two; it is also commonly called the **average** of the other two.

Thus, a is the arithmetic mean of $a - c$ and $a + c$.

If it be required to find the arithmetic mean between any two given quantities, as m and n, let x be the unknown mean, and we have
$$m - x = x - n;$$
whence $\quad 2x = m + n,$

and $\quad x = \dfrac{m+n}{2},$

which is the formula for the average of two quantities.

100. Harmonic progression. When three quantities are so related that the second is the arithmetic mean of the first and third, the reciprocals of these quantities constitute what is called a **harmonic progression**.

Suppose that b is an arithmetic mean between a and c.

Then
$$a - b = b - c,$$

and
$$1 = \frac{a-b}{b-c}.$$

Multiplying both members of this equation by $\frac{c}{a}$,
$$\frac{c}{a} = \frac{c}{a}\left(\frac{a-b}{b-c}\right),$$

which may be written
$$\frac{\frac{1}{a}}{\frac{1}{c}} = \frac{\frac{a-b}{a}}{\frac{b-c}{c}} = \frac{\frac{a-b}{ab}}{\frac{b-c}{bc}} = \frac{\frac{1}{b}-\frac{1}{a}}{\frac{1}{c}-\frac{1}{b}}.$$

If the reciprocals of a, b, c are respectively denoted by a', b', c', we now have
$$\frac{a'}{c'} = \frac{b'-a'}{c'-b'} = \frac{a'-b'}{b'-c'}.$$

This formula, $\frac{a'}{c'} = \frac{a'-b'}{b'-c'}$, is sometimes given as expressing the condition that three quantities, as a', b', c', shall be in harmonic progression.

101. Harmonic mean. In the last formula of the preceding article, b', the middle term, is called the **harmonic mean** between the other two quantities.

If any two quantities, as p and q, are given, and it is required to find x, the harmonic mean between the two, we have
$$\frac{1}{p} - \frac{1}{x} = \frac{1}{x} - \frac{1}{q};$$

PROGRESSIONS.

whence
$$\frac{2}{x} = \frac{1}{p} + \frac{1}{q} = \frac{p+q}{pq},$$

and therefore
$$x = \frac{2pq}{p+q}.$$

102. Geometric progression. A series is said to be a geometric series or **geometric progression** when the terms increase or decrease by a common factor.

Thus, $3 + 6 + 12 + 24 + \cdots$ is a geometric series, each term being obtained by multiplying the one before it by the factor 2.

Conversely, the constant factor may be found by dividing any term by the preceding one.

103. nth term. Considering the representative geometric progression,
$$a + ar + ar^2 + ar^3 + \cdots,$$
we see that in the third term the common factor is raised to the second power, in the fourth term it is raised to the third power, and so on. The nth term must therefore be $a(r^{n-1})$.

104. Sum of n terms. To find the sum of n terms in geometric progression, we may write
$$s = a + ar + ar^2 + ar^3 + \cdots + ar^{n-1};$$
$$rs = ar + ar^2 + ar^3 + \cdots + ar^{n-1} + ar^n.$$

Subtracting the first of these expressions from the second,
$$s(r-1) = -a + ar^n.$$
$$\therefore s = \frac{a(r^n - 1)}{r - 1} = \frac{a(1 - r^n)}{1 - r}$$
$$= \frac{a}{1-r} - \frac{ar^n}{1-r}.$$

Now if r is a proper fraction, the greater the value of n, the smaller is the value of r^n; and as neither a nor r changes in value, the fraction $\dfrac{ar^n}{1-r}$ may be made as small as we please by taking n sufficiently large. The entire expression $\dfrac{a}{1-r} - \dfrac{ar^n}{1-r}$ is then said to approach $\dfrac{a}{1-r}$ as its limiting value.

But the expression $\dfrac{a}{1-r} - \dfrac{ar^n}{1-r}$ is the sum of n terms; hence the sum of n terms approaches $\dfrac{a}{1-r}$ as its limit as more and more terms are included.

105. Geometric mean. When three quantities are in geometric progression, the middle one is said to be a **geometric mean** between the other two.

If we have a, b, c in geometric progression,

$$\frac{b}{a} = \frac{c}{b};$$

therefore $\qquad b^2 = ac,$

and hence $\qquad b = \pm \sqrt{ac}.$

It thus appears that a geometric mean and a mean proportional are the same thing.

CHAPTER XII.

INEQUALITIES.

106. Any quantity is said to be greater than another when the first quantity minus the other is a positive quantity.

If a represents the minuend quantity, and b the subtrahend, three cases arise:

a and b may both be positive;

one may be positive and the other negative;

both may be negative.

(1) If a and b are both positive quantities, and if $a - b$ is positive, $a > b$ by definition.

This is the case of common arithmetic.

(2) If a is any positive quantity, and b any negative quantity, $a - b$ is necessarily positive, and therefore $a > b$.

That is, any positive quantity is greater than any negative quantity.

Hence also, zero, the transition state (no magnitude) between positive and negative quantities, is greater than any negative quantity.

(3) Finally, of two negative quantities, the one which is numerically nearer to zero is greater than the other.

For example, $-3 > -7$ because $-3 - (-7) = 4$, a positive quantity.

107. If
$$a > b,$$
and c, any positive quantity, be added to each member of the inequality,
$$a + c > b + c,$$
for
$$(a + c) - (b + c) = a - b;$$
but $a - b$ is a positive quantity by the condition $a > b$.

Again, if c, any positive quantity, be subtracted from each member of the inequality,
$$a > b,$$
$$a - c > b - c,$$
because
$$(a - c) - (b - c) = a - b,$$
and $a - b$ is a positive quantity.

Hence, *if any positive quantity be added to or subtracted from each member of an equality, the inequality still holds good.*

108. If
$$a > b,$$
and each member be multiplied by any positive quantity, as c,
$$ca > cb;$$
for
$$ca - cb = c(a - b),$$
and both c and $a - b$ are positive by the conditions stated.

If
$$a > b,$$
and each member be divided by any positive quantity, as c,
$$\frac{a}{c} > \frac{b}{c},$$
for
$$\frac{a}{c} - \frac{b}{c} = \frac{1}{c}(a - b),$$
and each factor of this last expression is positive.

INEQUALITIES.

Hence, *if each member of an inequality be multiplied or divided by any positive quantity, the inequality still holds good.*

109. If $a > b,$

and $-c$, any negative quantity, be added to each member,

$$a + (-c) > b + (-c),$$

for $\quad [a + (-c)] - [b + (-c)] = a - b.$

The case is evidently identical with that of subtracting the positive quantity c from each member of the equation.

Similarly, subtracting any negative quantity, as $-c$, from each member of the inequality is the same as adding the positive quantity c to each member.

Hence, *if any negative quantity be added to or subtracted from each member of an inequality, the inequality still holds good.*

It follows that a term may be transposed from one member to another if its sign be changed;

for if $\quad a > b + c,$

adding $-c$ to each member to the inequality,

$$a - c > b.$$

110. If $a > b,$

and each member be multiplied by any negative quantity, as $-c$,

$$-ca < -cb,$$

for $\quad -ca - (-cb) = -c(a-b);$

and since $a - b$ is positive, $-c(a-b)$ is negative.

Again, if $\quad a > b,$

and each member be divided by any negative quantity as $-c$,
$$\frac{a}{-c} < \frac{b}{-c},$$
for
$$\frac{a}{-c} - \frac{b}{-c} = \frac{1}{-c}(a-b);$$
and this last expression is a negative quantity.

Hence, *if each member of an inequality be multiplied or divided by any negative quantity, the result is an inequality with the sign reversed.*

As a special case, suppose $-c$ is -1,
then if
$$a > b,$$
$$-a < -b;$$
that is, changing the sign of each member of an inequality reverses the sign of inequality.

111. If two inequalities,
$$a_1 > b_1,$$
$$a_2 > b_2,$$
are given and it is required to find the relation of
$$a_1 + a_2 \text{ to } b_1 + b_2,$$
let $\quad a_2 = a_1 + m$ and $b_2 = b_1 + n$;
then $\quad a_1 + a_2 = 2a_1 + m$ and $b_1 + b_2 = 2b_1 + n$;
hence $(a_1 + a_2) - (b_1 + b_2)$ becomes $2(a_1 - b_1) + (m - n)$,
in which $a_1 - b_1$, and hence $2(a_1 - b_1)$, is positive in the condition;
now since
$$a_1 + m > b_1 + n,$$
$$(a_1 - b_1) + (m - n) \text{ is positive};$$

INEQUALITIES.

and therefore
$$2(a_1 - b_1) + (m - n) \text{ is positive;}$$
hence $\quad a_1 + a_2 > b_1 + b_2.$

This operation can evidently be extended to any number of inequalities.

Hence, *the sum of all the greater members of a series of inequalities is greater than the sum of all the less members.*

112. If $\quad a_1 > b_1 \text{ and } a_2 < b_2,$
we can reach no conclusion in regard to the relation of
$$a_1 + a_2 \text{ and } b_1 + b_2.$$

Using the notation of the preceding article, $a_2 < b_2$ becomes $a_1 + m < b_1 + n$;

hence $\quad a_1 - b_1 < n - m;$

that is, $\quad n - m > a_1 - b_1;$

and since $a_1 - b_1$ is positive by the conditions given, $n - m$ must also be positive.

Combining the given inequalities as proposed, we have as new members,
$$2a_1 + m \text{ and } 2b_1 + n;$$
but $\quad (2a_1 + m) - (2b_1 + n),$

that is, $\quad 2(a_1 - b_1) + (m - n),$

or $\quad 2(a_1 - b_1) - (n - m),$

is of unknown sign; for, although, as we have seen, $n - m$ is positive and greater than $a_1 - b_1$, we do not know whether it is greater or less than $2(a_1 - b_1)$; it will be greater in some cases and less in others.

For example, if we have the inequalities,
$$7 > 5 \text{ and } 4 < 9,$$
the sum of the first members is at once seen to be less than the sum of the second members; and if we use the formula $2(a_1 - b_1) - (n - m)$, we ought to find
$$n - m > 2(a_1 - b_1),$$
and hence $2(a_1 - b_1) - (n - m)$, a negative quantity.

$a_1 = 7, \ b_1 = 5,$
$a_2 = a_1 + m = 4 = 7 + m, \ \therefore \ m = -3;$
$b_2 = b_1 + n = 9 = 5 + n, \ \therefore \ n = 4;$
hence $n - m = 7,$
and $2(a_1 - b_1) - (n - m) = 4 - 7 =$ a negative quantity.

Again, from the inequalities,
$$11 > 8 \text{ and } 3 < 4,$$
$$m = -8 \text{ and } n = -4;$$
hence $2(a_1 - b_1) - (n - m) = 6 - 4 =$ a positive quantity.

113. If $\qquad a_1 > b_1$
and $\qquad a_2 > b_2,$
and the four quantities, a_1, b_1, a_2, b_2 are all positive,
$$a_1 a_2 > b_1 b_2;$$
for suppose $a_2 = b_2 + c$, in which c is necessarily positive; if we multiply the first member of the first inequality by a_2, and the second member by $b_2 + c$, we have
$$a_1 a_2 > b_1 b_2 + b_1 c; \qquad \text{(Art. 108)}$$
hence $a_1 a_2 - b_1 b_2 - b_1 c$ is a positive quantity

INEQUALITIES.

in which the three terms a_1a_2, b_1b_2, b_1c are individually positive;

hence $a_1a_2 - b_1b_2$ is positive,

and therefore $a_1a_2 > b_1b_2$.

It is left to the student to examine the case in which the four quantities a_1, b_1, a_2, b_2 are negative, and also the case in which the members of one of the inequalities are positive and the members of the other inequality negative.

114. If we have any number of inequalities as $a_1 > b_1$, $a_2 > b_2$, $a_3 > b_3$, $\cdots a_n > b_n$, in which a_1, a_2, a_3, $\cdots a_n$ and b_1, b_2, b_3, $\cdots b_n$ are positive quantities, then by a repetition of the process of the preceding article,

$$a_1a_2a_3 \cdots a_n > b_1b_2b_3 \cdots b_n.$$

If $a_1 = a_2 = a_3 = \cdots = a_n = a,$

and $b_1 = b_2 = b_3 = \cdots = b_n = b,$

we have, from the inequality just obtained,

$$a^n > b^n,$$

when $a > b$ and a and b are positive.

EXAMPLES.

1. If $mn > 1$, and m is a proper fraction, what is the relation of n to the reciprocal of m?

2. If $a^2 > b^2$, what is the relation of a to b?

3. Show that $a^2 + b^2 > 2ab$.

4. Which is the greater, $\dfrac{p+q}{2}$ or $\dfrac{2pq}{p+q}$?

5. If a is any real positive quantity, show that

$$a + \frac{1}{a} \gtreqqless 2.$$

6. If a, b, c are positive quantities,

$$a^2 + b^2 + c^2 > bc + ca + ab.$$

7. If $\qquad a > b,$

what relation holds between a^{-n} and b^{-n}, a, b, n being any positive quantities.

8. If $\qquad a_1 > b_1$ and $a_2 > b_2,$

examine the threefold statement

$$a_1 - a_2 \gtreqless b_1 - b_2.$$

Can two of these signs of relation be ruled out of the formula?

9. If $\qquad a_1 > b_1$ and $a_2 > b_2,$

does one of the signs of relation hold, to the exclusion of the others, in the expression

$$\frac{a_1}{a_2} \gtreqless \frac{b_1}{b_2}.$$

CHAPTER XIII.

VARIATION.

115. One quantity is said to **vary directly** as another quantity when the two quantities are so connected that if one is changed the other changes in the same ratio.

Suppose that the quantities are a and b, and that when a has the values $a_1, a_2, a_3, \cdots$, b has corresponding values $b_1, b_2, b_3, \cdots$.

By definition,
$$\frac{a}{b}=\frac{a_1}{b_1},\ \frac{a}{b}=\frac{a_2}{b_2},\ \frac{a}{b}=\frac{a_3}{b_3},\ \cdots;$$

hence $$\frac{a_1}{b_1}=\frac{a_2}{b_2}=\frac{a_3}{b_3}=\cdots=k,$$

in which k is the value of any one of the ratios.

Then $\quad a_1 = kb_1,\ a_2 = kb_2,\ a_3 = kb_3,\ \cdots$.

It thus appears that
if $\quad a \propto b$,
$\quad a = b$ multiplied by some constant factor.

It follows that, if the conditions of a problem enable us to state a variation relation, the introduction of an undetermined constant factor enables us to pass from the variation relation to an equational relation. The value of this factor must then be determined from the data of the problem.

For example, we know from geometry that the circumference of a circle varies as the radius. If c be the circumference and r the radius, we have

$$c \propto r;$$

then $\qquad c = kr.$

If we make r equal to unity, we discover the meaning and value of k: it is the circumference of a circle whose radius is unity; and from geometry we know that the circumference of such a circle is 2π;

that is, $\qquad c = 2\pi r.$

116. One quantity a is said to **vary inversely** as another quantity b when a varies directly as the reciprocal of b.

That is, $\qquad a \propto \dfrac{1}{b},$

and therefore $\qquad a = \dfrac{k}{b}.$

For example, if the compression of a gas be so conducted that its temperature remains constant, the volume of the gas varies inversely as the pressure to which it is subjected. If v represents the volume and p the pressure,

$$v \propto \dfrac{1}{p}$$

and $\qquad v = \dfrac{k}{p};$

hence also, $\qquad p \propto \dfrac{1}{v}.$

117. If any quantity a depends upon two quantities b and c, varying as b when c is constant and varying as c when b is constant, then a varies as the product of b and c when b and c vary together.

Suppose, first, that a and b vary together, c remaining constant; and let the simultaneous values of the three quantities be a_1, b_1, c; then suppose that a_1 and c vary together, b_1 remaining constant, and let the three quantities be indicated by a_2, b_1, c_1; we have

$$\frac{a}{a_1} = \frac{b}{b_1}$$

and

$$\frac{a_1}{a_2} = \frac{c}{c_1}.$$

Multiplying the corresponding members of these expressions together,

$$\frac{a}{a_1} \times \frac{a_1}{a_2} = \frac{b}{b_1} \times \frac{c}{c_1};$$

that is,

$$\frac{a}{a_2} = \frac{bc}{b_1 c_1},$$

and hence,

$$a = \left(\frac{a_2}{b_1 c_1}\right) bc;$$

but a_2, b_1, c_1 are particular states or values of the varying quantities a, b, c; hence the expression $\frac{a_2}{b_1 c_1}$ is a constant, and therefore $\quad a \propto bc$.

For example, the stress between two gravitating bodies varies as the product of their masses and inversely as the square of the distance between the bodies; hence, if F represents the stress, or force of attraction, between the bodies, and m and m' represent their masses, and r is the distance between them,

$$F \propto \frac{mm'}{r^2}.$$

EXAMPLES.

1. If $x \propto \dfrac{1}{y}$ and $y \propto \dfrac{1}{z}$, show that $x \propto z$.

2. If $a \propto x$ and $x \propto y$, express a relation between a and y.

3. If $a \propto b$, show that $a^n \propto b^n$.

CHAPTER XIV.

REVIEW.

118. An Introduction to Algebra may well include the following words of general advice given by Professor Chrystal. The advice is equally pertinent to what has preceded this chapter and to what is to follow in Part II.

"Never make a step that you cannot justify by reference to the fundamental laws of algebra. In other respects make the freest use of your judgment as to the order and arrangement of steps.

"Take the earliest opportunity of getting rid of redundant members of a function, unless you see some direct reason to the contrary.

"Cultivate the use of brackets as a means of keeping composite parts of a function together, and do not expand such brackets until you see that something is likely to be gained thereby, inasmuch as it may turn out that the whole bracket is a redundant member, in which case the labor of expanding is thrown away, and merely increases the risk of error.

"Take a good look at each part of a composite expression, and be guided in your treatment by its construction; *e.g.* by the factors you can perceive it to contain, by its degree, and so on.

"Avoid the unthinking use of mere rules as much as possible, and use instead processes of inspection and general principles. In other words, use the head rather

106   ALGEBRA.

than the fingers. But if you do use a rule involving mechanical calculation, be patient, accurate, and systematically neat in the working. It is well known to mathematical teachers that quite half the failures in algebraical exercises arise from arithmetical inaccuracy and slovenly arrangement.

"Make every use you can of general ideas, such as homogeneity and symmetry, to shorten work, to foretell results without labour, and to control results and avoid errors of the grosser kind." *

EXAMPLES FOR REVIEW.

1. Show that the sum of two linear expressions will in general be linear.

2. Give an example in which the sum of two or more linear expressions is of zero dimensions.

3. If a is a negative quantity and b a positive quantity, what is the sign of the expression
$$(-1)^n (a)^n (b)^n,$$
(1) when n is odd, (2) when n is even?

4. If a is negative and b is positive, what is the sign of the expression
$$(-1)^{n+1}(-a)^{\frac{n}{2}}(b)^{-n},$$
(1) when n is odd, (2) when n is even?

5. (a) Prove that the product of two homogeneous expressions is itself a homogeneous expression.

(b) If one of the factors is of m dimensions and the other of n dimensions, the product will be of what dimensions?

---

* Chrystal's *Algebra*, Part I, p. 143.

REVIEW.

6. Give a definition of a *reciprocal* of a quantity as implied in the context of chapter V.

7. What is the reciprocal of the reciprocal of a quantity?

8. What operations are indicated by $\dfrac{(a^z + x^a)^2}{\sqrt{a^z x^a}}$?

9. Analyze the statement
$$(a + b)^3 = a^3 + 3\,a^2 b + 3\,ab^2 + b^3,$$
with reference to law and convention.

10. What is the value of a quantity if it is equal to its own square root?

11. Write the square of the binomial $a^{\sqrt{2}} - a^{-\sqrt{2}}$.

12. If $4x^2 + y^2 = 4xy$, find the ratio of x to y.

13. If $a : b = c : d$, show that $\dfrac{a-b}{a-2b} = \dfrac{c-d}{c-2d}$.

14. Given $\sqrt{x} + \sqrt{5x+1} = 1$. State the degree of this equation, and give a consistent definition of 'degree.'

15. Perform the operation indicated by
$$(2\,x^{\frac{1}{3}} - by^{-2})^2.$$

16. Given $\dfrac{\sqrt{x+1} - \sqrt{x-1}}{\sqrt{x+1} + \sqrt{x-1}} = \dfrac{x}{2}$. Find the value of x.

17. Find the sum of the series
$$1 + \tfrac{1}{2} + \tfrac{1}{4} + \cdots \text{ to 6 terms.}$$

18. Find an expression for x from the equation
$$x^a + 2\,x^{-a} - 3 = 0.$$

19. Simplify $\dfrac{\frac{1}{3}\sqrt{27} + \frac{1}{2}\sqrt{\frac{1}{12}}}{\sqrt[3]{3}}(\sqrt{-1})^6$.

20. What is the degree of the equation
$$y = \frac{x}{1 + x^2},$$
viewed (1) as an equation in x, (2) as an equation in y, (3) as an equation in x and y?

21. What is the degree of the equation
$$y^2 = \frac{x^3}{2a - x},$$
viewed (1) as an equation in x, (2) as an equation in y, (3) as an equation in x and y.

22. If $\dfrac{x^2 + 1}{x} = z$, write the value of x in terms of z.

23. The product of the two roots of a quadratic equation is 12, and the sum of the roots is 8; what is the equation?

24. Write down the roots of
$$x^2 - (a + 1)x + (a^2 - 1) = 0.$$

25. If p equals zero in the equation
$$x^2 + px + q = 0,$$
how must q be limited in order that the roots may be real?

26. If the area of a plane triangle varies as the height when the base is constant, and varies as the base when the height is constant, show that the area varies as the product of the base and height.

27. If the volume v of a sphere varies as r the radius and as the area of a cross-section through the centre, and the cross-section varies as the square of the radius, show that
$$v \propto r^3.$$

Books of reference recommended for teachers of elementary algebra.

BALL, W. W. R.   *A Short History of Mathematics.*
CAJORI, F.   *A History of Elementary Mathematics.*
CHRYSTAL, G.   *Algebra.*
CLIFFORD, W. K.   *The Common Sense of the Exact Sciences.*
CONANT, L. L.   *The Number Concept.*
MCLELLAN, J A., AND DEWEY, J   *The Psychology of Number.*
PEARSON, K.   *The Grammar of Science.*

PART II.

CHAPTER XV.

DERIVATIVES.

119. In the definition of direct variation given in Art. 115, two things are implied:

(1) quantities may be viewed as *varying in value;*

(2) quantities may be so connected that if one varies in value, the other necessarily varies in value.

When one quantity thus depends on another for its value, the former is said to be a **function** of the latter.

Ex. 1. $2\pi r$, the circumference of a circle, changes in value as r changes; it is therefore a function of r.

Ex. 2. If the volume of a gas varies inversely as the pressure, that is,

if $$v = \frac{k}{p},$$ (Art. 116)

v is a function of p, increasing in value as p diminishes, and decreasing as p increases.

Ex. 3. If $$F \propto \frac{mm'}{r^2},$$ (Art. 116)

and the masses of the two bodies do not change, F the stress, or gravitation force, between them is a function of r, the distance between the bodies.

If x represents the fundamental quantity or **independent variable**, the usual symbol for the **dependent variable** or function of x is $f(x)$. To distinguish one function from another, we use similar symbols, as $\phi(x)$, $F(x)$, $f_1(x)$, etc.

It should be carefully noticed that a new convention is now introduced The parentheses merely serve to separate the quantity symbol x from the other symbol f, ϕ, F, etc., which is not a quantity symbol, and hence not a factor. $f(x)$, $\phi(x)$, etc., is only algebraic shorthand for the expression, 'a function of the varying quantity x.'

It is also to be noticed that the term 'function' might have been used in many connections in which the term 'expression' was used in Part I.

Thus the expression $ax + b$ is evidently a function of x; its value must change if x changes, and this is the test of one quantity's being a function of another.

120. The term **constant** has already been used to denote a quantity which is viewed as not changing in value in a given expression. Two kinds of constants, literal and numerical, have also been noticed (Art. 75). To a literal or general constant we shall now apply the term **arbitrary**, because an arbitrary value may be assigned to it. All other constants will be classed as **absolute**.

Thus the expression $ax + b$ contains the arbitrary or general constants a and b, while the expression $4x - 2$ contains the absolute or particular constants 4 and -2.

Again, if $f(x) = \dfrac{4n}{3}\pi x^3$, 4, 3, and π are absolute constants, n is arbitrary and independent of x.

121. Two modes of variation present themselves. (1) The number of roses in a handful may be varied by

DERIVATIVES. 113

adding one, and another, and another, until the number has changed from a to b. Or, we may add several at a time until the number has changed from a to b. But we cannot do less than add one whole rose at a time; for, in this case, the variation element is a whole unit; that is, a whole rose, and not any fraction of it. Again, in measuring the quantity of wheat in a bin, the variation element was taken in Art. 1 as one whole bushel, the chosen unit of measure; and the quantity of wheat in the bin varied, becoming less by a bushel at a time.

(2) Suppose now that the fifty bushels of wheat are ground to the finest flour, and the flour allowed to run through a very small hole; we have a rude example of another kind of variation. The quantity of wheat is diminishing in this case also; but the variation element is so small that we say the variation is **continuous**.

To illustrate further, we may measure a day with a minute as a unit of measure, and say that a day contains 1440 minutes; but this is only an artificial convenience. Time does not increase a minute at a time, or even a second at a time, but by elements of time which are immeasurably small fractions of a second.

By the term **variable** we shall mean a quantity which changes in the second manner described: not by jumps or finite amounts, but by indefinitely small amounts.

122. Functions may be classified as **algebraic** and **transcendental**.

Algebraic functions are those which involve only the six operations, — addition, subtraction, multiplication, division, involution, and evolution, the exponents indicating the last two operations being constant.

The consideration of transcendental functions must be

preceded by a study of trigonometry and the theory of logarithms. Until the chapter on the theory of logarithms is reached, it is to be understood that $f(x)$ denotes only algebraic functions.

A **rational algebraic function** of a quantity is one in which the quantity is free from fractional exponents (Art. 48).

An **integral algebraic function** of a quantity is one in which the quantity is free from negative exponents (Art. 65).

The expression

$$p_0 x^n + p_1 x^{n-1} + p_2 x^{n-2} + \cdots + p_{n-1} x + p_n$$

is a rational integral algebraic function of x, n being positive and integral, and the coefficients $p_0, p_1, \cdots p_n$, being independent of x. The limitations in regard to the exponents are to be understood to apply only to the exponents of x; the coefficients p_0, p_1, etc., may have negative and fractional exponents.

Whenever the above expression is used in the following pages, n is positive and integral unless the contrary is expressly stated; also p_0, p_1, etc., are real quantities.

It is to be further observed that this expression is non-homogeneous and of the nth degree, and that it contains $n + 1$ terms if none of the constants $p_0, p_1, \cdots p_n$ are zero.

123. As x in $f(x)$ changes, it is obvious that the value of $f(x)$ must change also, whatever kind of function it may be.

For example, let $f(x) = \frac{4}{3}\pi x^3$; if we assign the value 3 inches to x, the volume of the sphere is 36π cubic inches; if the radius is 4 inches, the volume is $85\frac{1}{3}\pi$ cubic inches.

Suppose x to change by taking an increment h, so that $f(x)$ becomes $f(x+h)$. If the first value of the function be subtracted from the second, the remainder
$$f(x+h) - f(x)$$
is the increment of the function due to the increment of the variable, and $\dfrac{f(x+h) - f(x)}{h}$ is the ratio of the increment of the function to that of the variable.

Let h now be supposed to diminish without limit, *i.e.* to decrease until it differs from zero by less than any assignable quantity. The value of the ratio $\dfrac{f(x+h) - f(x)}{h}$, when this supposition is made regarding h, is represented by $f'(x)$, and is called the **first derived function of $f(x)$** (or, briefly, the **first derivative**), since it is derived from $f(x)$, and is itself, in general, some function of x. When h diminishes without limit, $f(x+h) - f(x)$ also diminishes without limit, so that in $f'(x)$ we have the ratio of two infinitely small quantities; but the ratio itself is, in general, some finite quantity.

In case the function is such that any real finite value of x renders $f(x)$ imaginary or impossible, $f'(x)$ also becomes imaginary or impossible. For such a value of x, $f(x)$ is said to be discontinuous.

To illustrate the nature of $\dfrac{f(x+h) - f(x)}{h}$, let $f(x) = 3x^2$.

Then we have, when x takes an increment,
$$f(x+h) = 3(x+h)^2 = 3(x^2 + 2xh + h^2),$$
and
$$\frac{f(x+h) - f(x)}{h} = \frac{3(x^2 + 2xh + h^2) - 3x^2}{h} = 6x + 3h;$$

116   ALGEBRA.

but $6x + 3h = 6x$ when h is made infinitely small.
$$\therefore f'(x) = 6x.$$

124. We have now to find expressions for the first derivative of a function when the given function is composed of other functions of the fundamental variable.

We begin with a function which consists of the algebraic sum of two functions.

If $\qquad f(x) = f_1(x) + f_2(x),$

and x takes the increment h, we have

$$\frac{f(x+h) - f(x)}{h} = \frac{f_1(x+h) + f_2(x+h) - f_1(x) - f_2(x)}{h}$$

$$= \frac{f_1(x+h) - f_1(x)}{h} + \frac{f_2(x+h) - f_2(x)}{h};$$

and in the limit $f'(x) = f_1'(x) + f_2'(x);$ \hfill (1)

hence, *the derivative of the sum of two functions is the sum of the derivatives of the functions.*

It is evident that the same proof would apply to any number of functions connected by plus and minus signs.

A constant, because it is a constant, has no increment; and if we attempt to express its derivative we have nothing to divide by h. This amounts to saying that *the derivative of a constant is zero.*

125. Let the given function consist of the product of two functions, as expressed by

$$f(x) = f_1(x) f_2(x);$$

what is the expression for $f'(x)$?

Let $u = f_1(x)$, and $u + h_1 = f_1(x + h)$, h_1 being the increment of u due to the increment h, which x has taken; also let $v = f_2(x)$, and $v + h_2 = f_2(x + h)$.

Then $f(x+h) = f_1(x+h)f_2(x+h) = (u+h_1)(v+h_2)$,

and $\quad f(x+h) - f(x) = (u+h_1)(v+h_2) - uv$

$$= vh_1 + uh_2 + h_1h_2;$$

hence $\quad\dfrac{f(x+h)-f(x)}{h} = \dfrac{vh_1}{h} + \dfrac{uh_2}{h} + \dfrac{h_1h_2}{h};$

but when h diminishes without limit,

$$\frac{h_1}{h} = f_1'(x), \text{ and } \frac{h_2}{h} = f_2'(x),$$

and the term $\dfrac{h_1h_2}{h}$ is disposed of by writing it in the form $h_1\left(\dfrac{h_2}{h}\right)$ and observing that as h_1 diminishes without limit, any quantity (except ∞) multiplied by h_1 diminishes without limit and is therefore dropped.

Hence we have

$$\begin{aligned}f'(x) &= vf_1'(x) + uf_2'(x) \\ &= f_2(x)f_1'(x) + f_1(x)f_2'(x);\end{aligned} \qquad (2)$$

and this result when translated becomes the theorem, *the derivative of the product of two functions is the derivative of the first function multiplied by the second function, plus the derivative of the second function multiplied by the first function.*

In a similar manner, we may find the derivative of the product of three or more functions of x.

Thus if

$$f(x) = f_1(x)f_2(x)f_3(x),$$
$$f'(x) = f_2(x)f_3(x)f_1'(x) + f_1(x)f_3(x)f_2'(x) + f_1(x)f_2(x)f_3'(x).$$

126. By means of the result in the preceding article, we may now find the derivative of a function which

consists of some function affected with any positive integral exponent; that is, we may find $f'(x)$ when
$$f(x) = [\phi(x)]^n,$$
n being positive and integral.

Let $\qquad f(x) = f_1(x)f_2(x) \cdots f_n(x);$
then
$$\begin{aligned} f'(x) = {} & f_1'(x)f_2(x) \cdots f_n(x) \\ & + f_2'(x)f_1(x)f_3(x) \cdots f_n(x) + f_3'(x)f_1(x)f_2(x) \cdots f_n(x) \\ & + \cdots + f_n'(x)f_1(x)f_2(x) \cdots f_{n-1}(x), \end{aligned} \qquad (3)$$
a polynomial of x terms.

Now suppose
$$f_1(x) = f_2(x) = \cdots = f_n(x) = \phi(x);$$
then $\qquad f(x) = [\phi(x)]^n;$
and (3) becomes
$$f'(x) = n[\phi(x)]^{n-1}\phi'(x). \qquad (4)$$

The translation of formula (4) affords a rule for finding the derivative of $[\phi(x)]^n$.

One important special case requires notice; namely, the case in which $\phi(x)$ is simply x.

The derivative of the fundamental variable itself is seen to be unity, being $\dfrac{h}{h}$, hence

if $\qquad \phi(x) = x, \ \phi'(x) = 1;$
and therefore (4) becomes
$$f'(x) = nx^{n-1}. \qquad (5)$$

127. Formula (4) will now enable us to find the derivative of a function which consists of some function affected with any exponent whatever.

Suppose the exponent is a positive fraction, so that we have
$$f(x) = [\phi(x)]^{\frac{p}{q}},$$
p and q being positive and integral.

Raising $f(x)$ to the qth power, and also raising to the qth power the expression to which $f(x)$ is equal,
$$[f(x)]^q = [\phi(x)]^p.$$

Now the *value* of an expression is independent of its *form* (compare Art. 72) It follows that two expressions or functions which are different in form but equal in value will receive the same increment when the fundamental variable common to the two functions takes an increment.

Hence, the *ratios* of these equal increments to the increment of the fundamental will be equal. That is, if two functions are equal, their derivatives are equal.

Applying (4) to each member of the equation
$$[f(x)]^q = [\phi(x)]^p,$$
and equating the resulting derivatives,
$$q[f(x)^{q-1}f'(x)] = p[\phi(x)]^{p-1}\phi'(x);$$
and solving for $f'(x)$,
$$f'(x) = \frac{p}{q} \frac{[\phi(x)]^{p-1}\phi'(x)}{f(x)^{q-1}}.$$

If, now, each of the two given equal functions, $f(x)$ and $[\phi(x)]^{\frac{p}{q}}$, are raised to the $(q-1)$th power, we have
$$[f(x)]^{q-1} = [\phi(x)]^{\frac{p}{q}(q-1)};$$
and, substituting the second member of this equation for its equal, the first member, which occurs in the denominator of the expression just obtained for $f'(x)$,

$$f'(x) = \frac{p[\phi(x)]^{p-1}\phi'(x)}{q[\phi(x)]^{\frac{p}{q}(q-1)}}$$

$$= \frac{p}{q}[\phi(x)]^{\frac{p}{q}-1}\phi'(x). \qquad (6)$$

128. If the function whose derivative is sought consists of some function affected with a negative integral exponent, we have
$$f(x) = [\phi(x)]^{-n},$$
n being integral.

Since
$$[\phi(x)]^{-n} = \frac{1}{[\phi(x)]^n},$$

$$f(x) = \frac{1}{[\phi(x)]^n},$$

and
$$f(x)[\phi(x)]^n = 1.$$

The derivative of the first member of this equation is found by applying the rule for the derivative of the product of two functions, together with the rule for the derivative of a power of a function.

The derivative of the second member is zero because that member is a constant. Equating the derivatives of the two members,
$$f'(x)[\phi(x)]^n + nf(x)[\phi(x)]^{n-1}\phi'(x) = 0;$$
solving for $f'(x)$, and eliminating $f(x)$ by means of the given equation, we have
$$f'(x) = -n[\phi(x)]^{-n-1}\phi'(x). \qquad (7)$$

129. Finally, if the case is that of a function affected with a negative fractional exponent, we write
$$f(x) = [\phi(x)]^{-\frac{p}{q}},$$
p and q being individually integral and positive.

DERIVATIVES.

Then $$f(x) = \frac{1}{[\phi(x)]^{\frac{p}{q}}},$$

that is, $f(x)[\phi(x)]^{\frac{p}{q}} = 1;$

and hence, raising each member to the qth power,

$$[f(x)]^q[\phi(x)]^p = 1.$$

Proceeding as in Art. 128, we find that

$$f'(x) = -\frac{p}{q}[\phi(x)]^{-\frac{p}{q}-1}\phi'(x). \tag{8}$$

130. Comparing formulas 4, 6, 7, 8, it is seen that

if $$f(x) = [\phi(x)]^n,$$
$$f'(x) = n[\phi(x)]^{n-1}\phi'(x)$$

for all cases, that is, when n is positive and integral, positive and fractional, negative and integral, negative and fractional

The translation of the formula,

$$f'(x) = n[\phi(x)]^{n-1}\phi'(x),$$

therefore furnishes the only rule that is needed for finding the derivative of a function affected with any constant exponent.

It should be noticed, however, that since this expression for the first derivative contains $\phi'(x)$ as a factor, we may require various other rules if we are to find the expression for which $\phi'(x)$ is the symbol.

131. Since first derivatives are themselves functions of the fundamental variable, their first derivatives may be found. These last are called the **second derivatives** of the original function. This operation may be extended until we have third, fourth, fifth derivatives, etc. Second derivatives are usually indicated by the symbol $f''(x)$, and third derivatives by $f'''(x)$.

EXAMPLES.

1. Find the first derivatives of the following functions:
$2a^3x$; $\quad -\tfrac{1}{4}x^5$; $\quad \pi x^2$; $\quad 3x^4 - cx^2 + 1$; $\quad ax^5 + bx^4 + c$,
$-\tfrac{1}{2}x^3 + \tfrac{2}{3}x^0$; $\quad ax^{\tfrac{2}{3}}$; $\quad \dfrac{3x^{\tfrac{2}{3}}}{x^{\tfrac{1}{3}}}$; $\quad (x^2-1)(x+2)$.

2. Find also the second and third derivatives of these functions.

CHAPTER XVI.

RATIONAL INTEGRAL FUNCTIONS.

132. In this chapter it is proposed to establish certain theorems relating to the function,

$$p_0x^n + p_1x^{n-1} + p_2x^{n-2} + \cdots + p_{n-1}x + p_n,$$

described at the close of Art. 122.

We shall first inquire what this representative rational integral function becomes when x is increased by any quantity as h.

Let $\quad f(x) = p_0x^n + p_1x^{n-1} + \cdots + p_{n-1}x + p_n;$
then $\quad f(x + h) = p_0(x + h)^n + p_1(x + h)^{n-1} + \cdots$
$\qquad\qquad + p_{n-1}(x + h) + p_n.$

Developing each term by the binomial theorem (Art. 140), and arranging the result according to the ascending powers of h, we have

$p_0x^n + p_1x^{n-1} + \cdots + p_{n-1}x + p_n$
$\quad + h[np_0x^{n-1} + (n-1)p_1x^{n-2} + \cdots + 2p_{n-2}x + p_{n-1}]$
$\quad + \dfrac{h^2}{\underline{|2}}[n(n-1)p_0x^{n-2} + (n-1)(n-2)p_1x^{n-3} + \cdots + 2p_{n-2}]$
$\quad + \;\cdot\;\cdot\;\cdot\;\cdot\;\cdot\;\cdot\;\cdot\;\cdot\;\cdot\;\cdot\;\cdot\;\cdot\;\cdot\;\cdot$
$\quad + \dfrac{h^n}{\underline{|n}}*[n(n-1)\cdots(2)(1)p_0].$

* By the symbol $\underline{|n}$ is meant the product of the numbers, 1, 2, 3, $\cdots n$.
Thus $\underline{|5} = 1 \times 2 \times 3 \times 4 \times 5$, and is read: '*factorial five.*'

The polynomial written on the first line does not contain h; it is therefore the absolute term of the entire expression; it is remarkable also for being the original given function.

Further, the part written on the second line is composed of the factor h with a polynomial coefficient. Attention to this coefficient shows that it is the *first derivative* of the given function.

Similarly, the third line contains the second power of h with a polynomial coefficient which is seen to be the first derivative of the preceding coefficient, and which is therefore the *second derivative* of the given function.

These part polynomials are sufficient to indicate the composition of the omitted parts; each term will contain h raised to a power higher by one unit than the power of h in the preceding term, and each polynomial coefficient of h is the derivative of the polynomial of the preceding term. The factorial divisor accompanying h is also to be noted.

By this regular formation of the entire expression, it must contain $n + 1$ parts, and the last part or term will contain h^n together with the divisor $\lfloor n$ and a coefficient which is the nth derivative of the given function.

Having noticed the features of this expression for $f(x + h)$, we are now able to write it in a compact symbolic form, and we have

$$f(x+h) = f(x) + hf'(x) + \frac{h^2}{\lfloor 2} f''(x) + \frac{h^3}{\lfloor 3} f'''(x) + \cdots + \frac{h^n}{\lfloor n} f^n(x).$$

This formula is a particular case of a theorem known as **Taylor's theorem**.

133. Although the fundamental quantity represented by x varies continuously, as explained in Art. 121, and

RATIONAL INTEGRAL FUNCTIONS.

although any function of the fundamental necessarily varies with the fundamental, it does not follow that all functions vary continuously. Some functions* are of such a nature that when x varies from some specified value to another, the function becomes imaginary, or infinite, or impossible, or it changes abruptly from one value to another value differing from the first by a finite amount.

Now by means of the formula established in the preceding article, we can show that if the function is a rational integral algebraic function it varies continuously; that is, it changes by real and indefinitely small amounts just as the fundamental changes.

For we have

$$f(x+h)-f(x)= hf'(x)+\frac{h^2}{\lfloor 2}f''(x)+ \cdots +\frac{h^n}{\lfloor n}f^n(x),$$

in which the first member is the amount of change in the function when x changes by the amount h. This quantity $f(x+h)-f(x)$ will be possible, real, and finite if the second member is possible, real, and finite.

We therefore examine the second member upon the assumption that $f(x)$ is a rational integral function.

Considering the way in which first, second, and suc-

---

* After the student has made sufficient advancement in mathematics, he will find examples of such functions in studying the curves of the following equations:

$$\frac{x^2}{a^2}-\frac{y^2}{b^2}=1, \text{ the hyperbola};$$

$$y = \tan x, \text{ the tangent curve};$$

$$y = \log x, \text{ the logarithmic curve};$$

$$\frac{d^2V}{dr^2}+\frac{2}{r}\frac{dV}{dr}=-4\pi\rho, \text{ the curve of the potential.}$$

cessive derivatives are formed, it is evident that no derivative of a rational integral function can be either imaginary or infinite for any real finite value of x.

Neither is there any reason for saying that any of these derivatives are impossible in the sense in which we say that $\log x$ is impossible when x is any negative quantity.

Further, the expression as a whole cannot be infinite by virtue of being the summing of an infinite number of terms; for, since n is positive and integral, it follows that the expression consists of only a finite number of terms.

Finally, we notice that h occurs in each term, and occurs only in the numerator and with a positive exponent; hence, by supposing h to diminish, we can make the value of the second member as small as we please.

And thus we learn that the first member is an indefinitely small quantity when h becomes indefinitely small. This being the nature of the increment of the function, we know that the function itself is continuous.

Since $\dfrac{f(x+h)-f(x)}{h} = f'(x)$ when h diminishes without limit, if h is positive, $f(x+h)-f(x)$ and $f'(x)$ must have the same sign. Consequently, when $f'(x)$ is positive, $f(x)$ is increasing with x; and when $f'(x)$ is negative, $f(x)$ diminishes as x increases.

134. A function is said to be **evaluated** for any constant, absolute or arbitrary, when that constant replaces the variable wherever the variable occurs in the function.

If $f(x)$ is evaluated for any constant as a, we express the fact symbolically by $f(a)$ or by $f(x)]_a$.

If one of two differently formed but equal functions is

RATIONAL INTEGRAL FUNCTIONS. 127

evaluated for any constant, evidently the other must be evaluated for the same constant.

If a function contains an absolute term, the evaluation of the function for any constant cannot affect the value of that absolute term.

Suppose now that we have an equation whose members are two differently formed functions of the same variable, and one of the members contains an unknown or undetermined constant.

Let the equation be

$$f(x) = \phi(x, c).$$

The form in which the second member is written is meant to indicate that the function contains the unknown constant c.

It follows from what has been said above that we may evaluate $f(x)$ and $\phi(x, c)$ by assigning any chosen value to x, and the value of c will not be thereby affected.

Let us then assign some convenient known value to x, so that the equation reduces to an equation containing only known constants and the unknown constant c.

Solving this equation for c, we may now return to the original equation in x and replace c by its value which we have found.

Suppose that the data and conditions of a problem are such that b is a convenient and desirable quantity to use in evaluating. Then

$$f(b) = \phi(b, c),$$

and hence c equals an expression containing b and other known quantities. Let κ represent this expression.

Then the original equation becomes

$$f(x) = \phi(x, \kappa).$$

The principle above outlined is of such wide application in the mathematical sciences that the student should become thoroughly familiar with it.

135. If any rational integral function of the nth degree is divided by a linear function of the form $x - a$, the quotient, according to the principles of division, will also be a rational integral function, and will be of the $\overline{n-1}$th degree.

If we carry on the division until the remainder is of zero dimensions in x, what is the value of the remainder?

Let $f(x)$ represent the given function, $\phi(x)$ represent the quotient, and r represent the remainder. Then
$$f(x) = (x - a)\phi(x) + r.$$

According to the principle explained in Art. 134, we may assign any value we please to x without affecting the value of r, whatever it is, because r does not contain x.

If we let $x = a$, the equation becomes
$$f(a) = (a - a)\phi(a) + r,$$
but $(a - a)\phi(a) = 0\phi(a) = 0;$
hence $f(a) = r.$

It thus appears that *the remainder is the given function evaluated for* a.

The reason why we chose to evaluate for a rather than any other quantity is evident.

An important special case arises: that in which the given function is exactly divisible by the given linear binomial. Since r equals $f(a)$ in all cases, $f(a)$ now equals zero because r is zero.

Conversely, if $f(x)$ equals zero when x equals a, $f(x)$ is exactly divisible by $x - a$.

For we have
$$f(x) = (x-a)\phi(x) + r,$$
and $$f(a) = (a-a)\phi(a) + r = 0;$$
and since $(a-a)\phi(a)$ is zero,
r must be zero.

136. The preceding article suggests that any given rational integral function may be made to vanish, that is to equal zero, if it is evaluated for certain values.

The questions arise: How many such values may there be in any given case? Have they any special relation to the constants in the function, and if so, what is the relation?

In the present chapter we shall see what conclusions follow when the assumption is made that a rational integral function of the nth degree vanishes for more than n values of the variable.

Let $f(x) = p_0 x^n + p_1 x^{n-1} + p_2 x^{n-2} + \cdots + p_{n-1} x + p_n$,
and suppose that $f(x)$ vanishes when x is made equal to each of the values $a_1, a_2, a_3, \cdots a_n$, no two of these quantities being equal.

By the preceding article $f(x)$ is exactly divisible by $x - a_1$, and we have
$$f(x) = (x - a_1)(p_0 x^{n-1} + \cdots),$$
the second factor being of the $(n-1)$th degree.

Again, since $f(x)$ is divisible by $x - a_2$, and since $x - a_1$ and $x - a_2$ are prime to each other, the factor $(p_0 x^{n-1} + \cdots)$ must be divisible by $x - a_2$, and hence
$$f(x) = (x - a_1)(x - a_2)(p_0 x^{n-2} + \cdots).$$

Continuing the factoring in this manner, we have finally
$$f(x) = (x - a_1)(x - a_2) \cdots (x - a_n) p_0,$$

and $f(x)$ is thus resolved into n linear binomial factors with one other factor p_0.

Suppose now that $f(x)$ vanishes for a quantity b, this quantity not being equal to any of the quantities $a_1, a_2, \cdots a_n$.

Then $f(b) = (b - a_1)(b - a_2)(b - a_3) \cdots (b - a_n)p_0 = 0$.

Since none of the factors $b - a_1, b - a_2, \cdots b - a_n$ can equal zero, we must conclude that p_0 equals zero.

Continuing this process, it is shown that each one of the coefficients $p_0, p_1, \cdots p_n$ must equal zero.

Hence, *if a rational integral function of the nth degree vanishes for more than* n *values of the variable, the coefficient of each power of the variable must be zero.*

137. Let us now suppose that two rational integral functions of the nth degree are equal for more than n values of the variable.

Let $\quad p_0 x^n + p_1 x^{n-1} + \cdots + p_{n-1} x + p_n$
$\quad\quad = p'_0 x^n + p'_1 x^{n-1} + \cdots + p'_{n-1} x + p'_n,$

x having more than n values.

Transposing,

$$(p_0 - p'_0)x^n + (p_1 - p'_1)x^{n-1} + \cdots$$
$$+ (p_{n-1} - p'_{n-1})x + p_n - p'_n = 0,$$

and we have a function of the nth degree vanishing for more than n values of the variable. Therefore by the preceding article,

$$p_0 - p'_0 = 0, \ p_1 - p'_1 = 0, \cdots p_n - p'_n = 0;$$
$$i.e. \quad p_0 = p'_0, \quad\quad p_1 = p'_1, \text{ etc.}$$

RATIONAL INTEGRAL FUNCTIONS.   131

Hence the two expressions

$$p_0x^n + p_1x^{n-1} + \cdots + p_{n-1}x + p_n,$$
$$p'_0x^n + p'_1x^{n-1} + \cdots + p'_{n-1}x + p'_n,$$

are identical term for term, and the assumed equation reduces to such a statement as $\phi(x) = \phi(x)$, which holds true for any and all values of x.

Hence, *if two rational integral functions of the nth degree are equal for more than* n *values of the variable, they are equal for all values of the variable*

138. If the conclusion of the preceding article be taken as a condition and expressed,

if $p_0x^n + p_1x^{n-1} + \cdots + p_{n-1}x + p_n$
$$= p'_0x^n + p'_1x^{n-1} + \cdots + p'_{n-1}x + p'_n$$

for all values of x, *then it will follow that the coefficients of the like powers of* x *are equal and the expressions are identical.*

For, since the expressions are equal for all values of x, they are equal when x is zero; in that case, $p_n = p'_n$. Dropping p_n and p'_n and dividing by x,

$$p_0x^{n-1} + p_1x^{n-2} + \cdots p_{n-1} = p'_0x^{n-1} + p'_1x^{n-2} + \cdots p'_{n-1}.$$

Making x again equal to zero, $p_{n-1} = p'_{n-1}$; dropping these equal terms and dividing by x,

$$p_0x^{n-2} + p_1x^{n-3} + \cdots + p_{n-2} = p'_0x^{n-2} + p'_1x^{n-3} + \cdots + p'_{n-2}.$$

Continuing this operation, the coefficients of the like powers of x are found to be equal.

139. The proposition proved in Art. 138 is known as the **principle of undetermined coefficients**. The principle is much used for developing a function into a series in ascending powers of the variable.

In the preceding demonstration we notice that neither expression contains more than $n+1$ terms, a finite number; also, that we might have written

$$p_n + p_{n-1}x + \cdots + p_1 x^{n-1} + p_0 x^n$$
$$= p'_n + p'_{n-1}x + \cdots + p'_1 x^{n-1} + p'_0 x^n.$$

If, now, instead of a finite number of terms, we have the series

$$p_n + p_{n-1}x + p_{n-2}x^2 + \cdots = p'_n + p'_{n-1}x + p'_{n-2}x^2 + \cdots,$$

and this equality holds for all values of x, the demonstration would proceed as before, with the conclusion that the coefficients of the like powers of x are equal.

To illustrate the use of the principle, let it be required to develop $\frac{x}{1+x}$ into a series including the third power of x.

Let $\quad \frac{x}{1+x} = A + Bx + Cx^2 + Dx^3 + \cdots.$

Here we assume that the given function $\frac{x}{1+x}$ can be developed into a series in ascending powers of x; our problem is to discover what values the quantities A, B, C, etc., must have in order that the function may equal such a series.

Multiplying both members of the equality by $1+x$,

$$x = A + (A+B)x + (B+C)x^2 + (C+D)x^3 + \cdots.$$

The first member may be written

$$0 + x + 0\,x^2 + 0\,x^3 + \cdots.$$

RATIONAL INTEGRAL FUNCTIONS. 133

Equating the coefficients of the like powers of x,

$$A = 0, \ (A + B) = 1, \ \therefore \ B = 1;$$
$$B + C = 0, \ \therefore \ C = -1;$$
$$C + D = 0, \ \therefore \ D = 1.$$

The coefficients are now determined. Placing their values in the assumed series,

$$\frac{x}{1+x} = x - x^2 + x^3 - \cdots.$$

Ex. 2. Develop $\dfrac{1}{(1+x)^{\frac{1}{2}}}$.

Assume

$$\frac{1}{(1+x)^{\frac{1}{2}}} = A + Bx + Cx^2 + Dx^3 + Ex^4 + \cdots.$$

Squaring both members of this equality,

$$\frac{1}{1+x} = A^2 + AB \ \bigg| \ \begin{matrix} x + AC \\ AB \end{matrix} \ \bigg| \ \begin{matrix} x^2 + AD \\ B^2 \\ AC \end{matrix} \ \bigg| \ \begin{matrix} x^3 + AE \\ BC \\ BC \\ AD \end{matrix} \ \bigg| \ \begin{matrix} x^4 + \cdots; \\ BD \\ C^2 \\ BD \\ AE \end{matrix}$$

the quantities at the left of any vertical line being the terms of the complete polynomial coefficient of the power of x written at the right of that line.

Multiplying both members of the equality by $1 + x$,

$$1 = A^2 + AB \ \bigg| \ \begin{matrix} x + AC \\ AB \\ A^2 \end{matrix} \ \bigg| \ \begin{matrix} x^2 + \\ B^2 \\ AC \\ 2AB \end{matrix} \ \bigg| \ \begin{matrix} AD \\ BC \\ BC \\ AD \\ B^2 + 2AC \end{matrix} \ \bigg| \ \begin{matrix} x^3 + \\ \\ \\ \\ \end{matrix} \ \bigg| \ \begin{matrix} AE \\ BD \\ C^2 \\ BD \\ AE \\ 2AD + 2BC \end{matrix} \ \bigg| \ x^4 + \cdots.$$

Equating the coefficients of the like powers of x, as in the first example,

$$A^2 = 1,$$
$$2AB + A^2 = 0,$$
$$2AC + 2AB + B^2 = 0,$$
$$2AD + 2BC + 2AC + B^2 = 0,$$
$$2AD + 2BC + 2BD + 2AE + C^2 = 0.$$

Thus we have a set of five simultaneous equations from which to determine the values of the five unknown quantities involved. If these values be substituted in the assumed series,

$$\frac{1}{(1+x)^{\frac{1}{2}}} = 1 - \tfrac{1}{2}x + \tfrac{3}{8}x^2 - \tfrac{5}{16}x^3 + \tfrac{35}{128}x^4 - \cdots.$$

EXAMPLES.

1. Prove that a rational integral function evaluated for any real finite value, will be real and finite, as stated in Art. 133.

2. Show that if a rational integral function of the nth degree be divided by a rational integral function of the mth degree, the quotient will be a rational integral function of the $\overline{n-m}$th degree.

3. Show that the nth derivative of a rational integral function of the nth degree is a constant.

4. What would the remainder be if the function $x^5 - 2x^4 - x^3 + 3x^2 + x - 1$ were divided by $x - 2$?

5. The quantity c has such a value that $x^2 - 4x + 4$ is exactly divisible by $x - c$. Find the value of c by Art. 135.

6. Develop the following functions into series:

(3) $\dfrac{a}{b+cx}$;   (4) $\dfrac{1+x}{1+x+x^2}$;   (5) $\dfrac{1-3x}{1+x^2}$.

CHAPTER XVII.

BINOMIAL THEOREM.

140. In Art. 29 is given the full text of Clifford's translation of the statement

$$(a + b)^2 = a^2 + 2ab + b^2.$$

We have now to consider the corresponding expressions for $(a + b)^3$, $(a + b)^4$, etc., and to reach, if possible, a *general* form; that is, an expression for $(a + b)^m$ when the indicated operation is performed, m indicating that $a + b$ is multiplied by itself any number of times.

$$(a + b)^3 = (a + b)(a + b)^2 = (a + b)(a^2 + 2ab + b^2)$$
$$= a^3 + 3a^2b + 3ab^2 + b^3.$$

Proceeding in this way, by introducing the factor $a + b$ into the expression for any particular power of $a + b$, we have the expression for the next power.

Writing the series of expressions in order,

$(a + b)^2 = a^2 + 2ab + b^2,$

$(a + b)^3 = a^3 + 3a^2b + 3ab^2 + b^3,$

$(a + b)^4 = a^4 + 4a^3b + 6a^2b^2 + 4ab^3 + b^4,$

$(a + b)^5 = a^5 + 5a^4b + 10a^3b^2 + 10a^2b^3 + 5ab^4 + b^5,$

. . . .

$(a + b)^m = ?$

ALGEBRA.

An examination of these expressions shows that

1. each expression is *rational* and *integral* as regards both a and b;
2. it is *symmetrical* in a and b;
3. it is *homogeneous* in a and b;
4. it is a *complete* homogeneous expression; that is, every term is present which is consistent with homogeneity;
5. the terms are in descending powers of a and ascending powers of b;
6. as a result of 4, the number of terms is one greater than the number of units in the exponent of $a + b$;
7. the coefficient of the highest power of a is unity, and the coefficient of the second term is the exponent of $a + b$;
8. the coefficient of any particular term may be obtained by multiplying the coefficient of the preceding term by the exponent of a in that term and dividing it by the number which is one unit greater than the exponent of b in that term.

For example, the expression for $(a+b)^4$ may be written

$$a^4 + 4a^3b + \frac{4(3)}{2}a^2b^2 + \frac{4(3)(2)}{2(2+1)}ab^3 + \frac{4(3)(2)(1)}{2(2+1)(3+1)}b^4.$$

These characteristics of the forms obtained by actual multiplication now justify us in writing

$$(a+b)^m = a^m + ma^{m-1}b + \frac{m(m-1)}{\lfloor 2}a^{m-2}b^2$$
$$+ \frac{m(m-1)(m-2)}{\lfloor 3}a^{m-3}b^3$$
$$+ \cdot \cdot \cdot \cdot \cdot \cdot \cdot$$
$$+ \frac{m(m-1)(m-2)\cdots(m-\overline{m-1})}{\lfloor m}b^m.$$

BINOMIAL THEOREM.

The method of reasoning here employed is known as mathematical induction. It consists in comparing a number of *special* forms and observing their common characteristics, and thence inferring a *general* form.

The student of logic will notice that there is a fundamental difference between this method of arriving at a general proposition and the method of true induction as used in science.

141. In Art. 140 we have arrived at a formula which will enable us to write at once the expression for any power of any binomial without going through the work of multiplying.

In that formula m is positive and integral; we proceed now to derive an expression for the development of any binomial affected with any exponent whatever, positive or negative, integral or fractional.

Since there is no reason why one of the terms may not be a variable, we shall use the form $(a + x)^m$ instead of $(a + b)^m$ used above.

Our problem may then be stated:

To develop the function $(a + x)^m$ into a series, m being positive or negative, integral or fractional.

Let $(a + x)^m = A + Bx + Cx^2 + Dx^3 + Ex^4 + \cdots$,

in which A, B, C, etc., are coefficients to be determined. Finding the first, second, and successive derivatives of the given function, and also of the assumed series, we have

$$f'(x) = m(a + x)^{m-1}$$
$$= B + 2Cx + 3Dx^2 + 4Ex^3 + \cdots,$$
$$f''(x) = m(m - 1)(a + x)^{m-2}$$
$$= 2C + 2\cdot 3Dx + 3\cdot 4Ex^2 + \cdots,$$

$$f'''(x) = m(m-1)(m-2)(a+x)^{m-3}$$
$$= 2\cdot 3 D + 2\cdot 3\cdot 4 Ex + \cdots,$$
$$f^{iv}(x) = m(m-1)(m-2)(m-3)(a+x)^{m-4}$$
$$= 2\cdot 3\cdot 4 E + \cdots.$$

Since x is a variable, and these equations therefore true for all values of x, they are true when x equals zero; in that case we have

$$A = a^m;\quad B = ma^{m-1};\quad 2C = m(m-1)a^{m-2},$$

and $C = \dfrac{m(m-1)}{2} a^{m-2};\ 2\cdot 3 D = m(m-1)(m-2)a^{m-3}$,

and $D = \dfrac{m(m-1)(m-2)}{\underline{3}} a^{m-3}$; and so on.

Substituting in the assumed series the coefficients as thus determined,

$$(a+x)^m = a^m + ma^{m-1}x + \frac{m(m-1)}{\underline{2}} a^{m-2}x^2$$
$$+ \frac{m(m-1)(m-2)}{\underline{3}} a^{m-3}x^3$$
$$+ \frac{m(m-1)(m-2)(m-3)}{\underline{4}} a^{m-4}x^4 + \cdots.$$

This is the **binomial theorem**, of which the formula of Art. 140 is now seen to be merely a special case.

Comparison of several successive terms in this series shows us the law of the formation of each term so that we can write any specified term without first writing all the preceding ones.

Thus, the nth term is

$$\frac{m(m-1)(m-2)\cdots(m-(n-2))}{\underline{n-1}} a^{m-(n-1)}x^{n-1},$$

and the $(n+1)$th term is
$$\frac{m(m-1)(m-2)\cdots(m-(n-1))}{\underline{|n}}a^{m-n}x^n.$$

The student should now consider which of the characteristics numbered 1, 2, 3, 4, 5, 6, 7, 8, in Art. 140, hold good for the three cases:

> m fractional and positive;
> m fractional and negative;
> m integral and negative.

EXAMPLES.

1. Develop into series by the binomial theorem:
(1) $(a-x)^m$; (2) $(qa+rb)^m$; (3) $(qa+rx^n)^m$.

2. Describe the binomials given in Ex. 1, and by means of the series found in that example, develop the following functions to five terms:
$$(2-y)^{-2};\ (3a+2b)^{-3};\ (\tfrac{1}{2}l+x^2)^{\tfrac{2}{3}}.$$

3. How must the expression $(qa+rb)^m$ be limited in order that when developed it shall be symmetrical in a and b?

4. Develop the expressions given in Ex. 2, by the binomial formula directly; and compare the results with the results obtained in the manner prescribed in Ex. 2.

5. Write the 6th term of the development of $(4-x)^{-\tfrac{5}{2}}$.

6. Write the expression for the ratio of the $(n+1)$th term of the binomial formula to the nth term.

7. Translate
$$(a+x)^m = a^m + ma^{m-1}x + \frac{m(m-1)}{\underline{|2}}a^{m-2}x^2 + \cdots,$$
giving prominence to the fact that ascending powers of the variable appear and that a and m are constants.

CHAPTER XVIII.

CONVERGENCY OF SERIES.

142. The definition of a *series* has been given in Art. 95. If a series is of such a nature that it comes to no natural termination but consists of an unlimited number of terms, it is called an **infinite series**.

The student has discovered that only when m is positive and integral does the development of $(a + x)^m$ contain only a finite number of terms. The binomial theorem, the exponent of the binomial not being limited to positive integral values, becomes therefore an example of an infinite series.

An infinite series is said to be **convergent** when the sum of the first n terms cannot numerically exceed some finite quantity, however great n may be.

If by taking n large enough the series can be made numerically greater than any finite quantity, the series is said to be **divergent**.

Preliminary to the search for a test of convergency, we shall consider the geometrical series

$$1 + x + x^2 + x^3 + \cdots.$$

The sum of n terms of this series is $\dfrac{1-x^n}{1-x}$. Let $x < 1$; then as n increases, x^n diminishes, and by taking n sufficiently great we can make x^n as small as we please. Hence $\dfrac{1-x^n}{1-x}$ tends to the limit $\dfrac{1}{1-x}$ as n is increased.

CONVERGENCY OF SERIES.   141

Let $x>1$; then by taking n indefinitely large $\dfrac{1-x^n}{1-x}$, the sum of n terms, becomes indefinitely large. The given series is therefore convergent for values of x less than 1, and divergent for values greater than 1.

Similarly, the sum of n terms of the series
$$1-x+x^2-x^3+\cdots$$
is
$$\dfrac{1-(-1)^n x^n}{1+x}.$$

When $x<1$, the limit of the sum of this series is $\dfrac{1}{1+x}$, and the series is convergent as in the first case; but when $x>1$, x^n increases as n increases, and the sum of the series has no limit; *i.e.* the series is divergent.

143. The following proposition can now be established:
If, in a series of positive terms, as
$$\mu_1+\mu_2+\mu_3+\cdots+\mu_n+\mu_{n+1}+\cdots,$$
the ratio $\dfrac{\mu_{n+1}}{\mu_n}$ be less than a certain quantity itself less than unity, for all values of n *beyond a certain number, the series is convergent.*

Suppose k to be a fraction less than unity and greater than the greatest of the ratios

$$\dfrac{\mu_{n+1}}{\mu_n}\cdots\text{ beyond the number }n,$$

then
$$\dfrac{\mu_{n+1}}{\mu_n}<k, \quad \therefore \mu_{n+1}<k\mu_n; \qquad (1)$$

$$\dfrac{\mu_{n+2}}{\mu_{n+1}}<k, \quad \therefore \mu_{n+2}<k\mu_{n+1}; \qquad (2)$$

$$\dfrac{\mu_{n+3}}{\mu_{n+2}}<k, \quad \therefore \mu_{n+3}<k\mu_{n+2}; \qquad (3)$$

and so on.

Since $k\mu_n > \mu_{n+1}$, if we substitute $k\mu_n$ for μ_{n+1} in (2), we have $\mu_{n+2} < k^2\mu_n$. Making a similar substitution in (3), etc., and adding the corresponding terms of these inequalities,

$$(\mu_{n+1} + \mu_{n+2} + \mu_{n+3} + \cdots) < (k\mu_n + k^2\mu_n + k^3\mu_n + \cdots).$$

But since k is a proper fraction, $k + k^2 + k^3 + \cdots$ is a converging series with $\dfrac{k}{1-k}$ for its limit. The other factor μ_n decreases as n increases, tending to zero as its limit.

Hence the second member of the above inequality has zero as its limit. The first member of the inequality is the sum of the series after the nth term, and since it is less than $\dfrac{k\mu_n}{1-k}$, it must also have zero as its limit.

Now the sum of the n terms preceding $\mu_{n+1} + \mu_{n+2} + \cdots$ is a finite quantity, being the sum of a *finite* number of *finite* terms. Adding this to the series beginning with μ_{n+1}, the limit of the entire series is seen to be a finite quantity; hence the conclusion that the series is convergent.

On the other hand, if the ratio $\dfrac{\mu_{n+1}}{\mu_n}$ is greater than 1 for all values of n beyond a certain number, the series is divergent.

Proceeding as before, let $k > 1$; then

$$(\mu_{n+1} + \mu_{n+2} + \cdots) > \mu_n(k + k^2 + k^3 + \cdots).$$

Now both factors in the second member of this inequality increase without limit; consequently the first member, which is always greater than the second member, must increase without limit; the series is therefore divergent.

CONVERGENCY OF SERIES. 143

In the preceding argument we have assumed that the terms μ_1, μ_2, μ_3, etc., are all positive. The conclusions reached may be shown to hold in case the terms of the series are alternately positive and negative; for k now becomes negative, and the series will be convergent or divergent according as $-k$ is $<$ or >1.

144. The necessity of examining a series with reference to convergency rests in the fact that no practical use can be made of a diverging series.

If a series is found to be diverging, it is rejected for such values of the variable as render it diverging; or it is transformed into a series which converges, and, if possible, into one which converges rapidly, in order that only a few terms need be used.

An important example of such transformation will be found in the chapter on the theory of logarithms; and an example of a series that converges for all values of the variable, occurs in the theory of the construction of trigonometric tables.*

145. Since the binomial theorem covers so many cases of series, it is important to know what limitations must be imposed in order that the formula itself shall be converging.

---

* Hayes' *Elementary Trigonometry*, Chap. VII., Art. 82.

The series referred to is either one of the following:

$$\sin x = \frac{x}{1} - \frac{x^3}{\lfloor 3} + \frac{x^5}{\lfloor 5} - \frac{x^7}{\lfloor 7} + \cdots.$$

$$\cos x = 1 - \frac{x^2}{2} + \frac{x^4}{\lfloor 4} - \frac{x^6}{\lfloor 6} + \cdots$$

144   ALGEBRA.

To make the investigation, let $a=1$ in the formula. Then we have

$$(1+x)^m = 1 + mx + \frac{m(m-1)}{\underline{2}}x^2$$

$$+ \frac{m(m-1)(m-2)}{\underline{3}}x^3 + \cdots$$

$$+ \frac{m(m-1)\cdots(m-(n-2))}{\underline{n-1}}x^{n-1}$$

$$+ \frac{m(m-1)\cdots(m-(n-1))}{\underline{n}}x^n + \cdots.$$

Now the ratio $\frac{\mu_{n-1}}{\mu_n}$ is seen to be $\frac{m-n+1}{n}x$, which equals $\left(-1+\frac{m}{n}+\frac{1}{n}\right)x$.

The factor in parentheses tends to the limit -1 as n increases indefinitely. If $x<1$, the limit of the entire expression is therefore some proper fraction, and the series is convergent; but if $x>1$, the series is divergent.

Hence values of x greater than 1 are not admissible in the development of $(1+x)^m$.

The more general function $(a+x)^m$ may be written $a^m\left(1+\frac{x}{a}\right)^m$. Let $\frac{x}{a}=z$; applying the above results to $(1+z)^m$, z must be less than 1; otherwise the series would be divergent. But if z is less than 1, x is less than a.

Hence the binomial theorem, that is, the theorem expressing the development of $(a+x)^m$, does not hold when $x>a$.

CONVERGENCY OF SERIES.

EXAMPLES.

Determine whether the following series are convergent or divergent:

1. $\dfrac{x}{1\cdot 2}+\dfrac{x^2}{2\cdot 3}+\dfrac{x^3}{3\cdot 4}+\dfrac{x^4}{4\cdot 5}+\cdots.$

2. $1+\dfrac{2^2}{\underline{|2}}+\dfrac{3^2}{\underline{|3}}+\dfrac{4^2}{\underline{|4}}+\cdots.$

3. $\sqrt{\tfrac{1}{2}}+\sqrt{\tfrac{2}{3}}+\sqrt{\tfrac{3}{4}}+\sqrt{\tfrac{4}{5}}+\cdots.$

4. $1+3x+5x^2+7x^3+9x^4+\cdots.$

CHAPTER XIX.

THEORY OF EQUATIONS.

146. As defined in Art. 71, an equation is a statement in the language of algebra that two expressions (functions) are equal to each other.

If the functions are regarded (1) as functions of one quantity only, and if (2) they are differently constituted, each value of the quantity for which the statement is true is called a **root** of the equation.

If the function constituting the second member of the equation is transposed to the first member, so that the equation takes the form

$$F(x) - \phi(x) = f(x) = 0,$$

then, in accordance with Art. 74, a root of the equation is a value of x which causes $f(x)$ to vanish, so that the expression which is equal to itself is zero.

If condition (1) holds and (2) does not, the equation reduces to an **identity.**

For example, $x - 5 = x - 5$ is an identity, and the statement is true for *all* values of x.

If condition (2) holds and (1) does not, the equation is said to be **indeterminate**; and the statement is true for an indefinite number of sets of values of the quantities involved.

For example, the equation $2x - 3y + 1 = 0$ is indeterminate if viewed as an equation in x and y. We may assign to x all values that we please, and by properly

THEORY OF EQUATIONS.   147

assigning values to y, may keep the statement of equality true. Thus, if $x = 1$, y must equal 1; if $x = 2$, $y = \tfrac{5}{3}$; if $x = -1$, $y = -\tfrac{1}{3}$, etc.

In what follows, we shall consider equations in which the above stated conditions (1) and (2) hold, together with a third condition that the functions shall be rational integral algebraic functions; and our problem is: to discover the roots of such equations, — their number and character and values.

Except when otherwise defined, the symbol $f(x)$ will be used to denote the rational integral function
$$x^n + p_1 x^{n-1} + p_2 x^{n-2} + \cdots + p_{n-1} x + p_n.$$

147. Number of roots. In the chapter on equations, Part I, it has been shown that the general equation of the first degree has one root, and that the general equation of the second degree has two roots. We have now to inquire whether there is any special relation between the degree of the equation and the number of the roots which holds for any degree.

If the equation $f(x) = 0$ is of the nth degree, and it be admitted that some of the coefficients $p_1, p_2, p_3, \cdots p_n$ are not zero, it then follows from Art. 136 that $f(x)$ cannot vanish for more than n values of the variable; that is, the equation $f(x) = 0$ cannot have more than n roots.

Now let α_1 be a root, then $f(\alpha_1) = 0$; also, by a preceding article, $f(x)$ is exactly divisible by $x - \alpha_1$, and we have
$$f(x) = (x - \alpha_1) f_1(x) = 0,$$
in which the factor $f_1(x)$ is of the $(n-1)$th degree.

Dividing both members of this equation by $x - \alpha_1$, we have $f_1(x) = 0$.

Suppose a_2 is a root of $f_1(x) = 0$; then, as before,
$$f_1(x) = (x - a_2)f_2(x) = 0;$$
in which $f_2(x)$ is of the $(n - 2)$th degree.

Repeating this operation, we have finally an equation of the first degree, and
$$f(x) = (x - a_1)(x - a_2)(x - a_3) \cdots (x - a_n) = 0.$$

Thus the first member of the equation is resolvable into n linear binomial factors, and the second term of each binomial is a root of the equation.

Hence, *assuming that every equation of the form* $f(x)=0$ *has a root, real or imaginary, if the equation be of the* nth *degree, it has* n *roots and no more.*

Two or more of the roots may be equal; the binomial factors remain n in number, and we still speak of the equation as having n roots.

If the multiplication indicated by
$$(x - a_1)(x - a_2) \cdots (x - a_{n-1})(x - a_n)$$
be performed, it is evident that the term p_n of the original polynomial equals
$$(-1)^n a_1 a_2 a_3 \cdots a_{n-1} a_n.$$

Hence, p_n, *the absolute term in the equation* $f(x)=0$, *is exactly divisible by each root of the equation.*

148. Imaginary roots. If $f(x)$ is divisible by the product of any number of linear factors of the form $x - a$, it is, of course, divisible by any one of the individual linear factors.

Thus, if $f(x)$ is exactly divisible by $(x^2 - \beta^2)$, it is exactly divisible by $x - \beta$ and $x + \beta$ separately; and β and $-\beta$ are roots of the equation $f(x) = 0$.

Also, if $f(x)$ is exactly divisible by $(x-\alpha)^2-\beta^2$, it is exactly divisible by $x-(\alpha+\beta)$ and $x-(\alpha-\beta)$; hence, in this case $\alpha+\beta$ and $\alpha-\beta$ are roots of the equation $f(x) = 0$.

Also, if $f(x)$ is exactly divisible by $(x-\alpha)^2+\beta^2$, it is exactly divisible by $x-(\alpha+\beta\sqrt{-1})$ and $x-(\alpha-\beta\sqrt{-1})$; and in this case the equation $f(x) = 0$ has the conjugate imaginary roots $\alpha+\beta\sqrt{-1}$ and $\alpha-\beta\sqrt{-1}$.

It thus appears that *an equation whose coefficients are real may have imaginary roots; but these roots must occur in pairs, and must be conjugate,* for the product of a binomial of the form $x-(\alpha+\beta\sqrt{-1})$ by any binomial of the form $x-a$ will conduct to imaginary coefficients; and likewise, if $x-(\alpha+\beta\sqrt{-1})$ be multiplied by any binomial of the form $x-(\gamma+\delta\sqrt{-1})$ (in which γ does not equal α and δ does not equal $-\beta$), imaginary coefficients will result.

149. Irrational roots. By a proof similar to the one given in the preceding article, it may be shown that though the coefficients of an equation are rational, it may have surd roots; but these surd roots, if occurring at all, must occur in pairs, and must be the correlated forms $\alpha+\sqrt{\beta}$ and $\alpha-\sqrt{\beta}$.

EXAMPLES.

1. Form the cubic equations whose roots are:

  (1) -1, 2, 5;

  (2) $-\tfrac{2}{3}$, $\tfrac{3}{4}$, 4;

  (3) a, -3, 3.

2. Form with real coefficients the equations of the fourth degree whose roots are:

(1) $4\sqrt{-1}$, 1, -1;

(2) $2-\sqrt{-3}$, $\frac{1}{2}$, -2;

(3) $p\sqrt{-1}$, $q\sqrt{-1}$.

3. Solve the equation
$$x^3 - 8x^2 + 6x + 52 = 0,$$
one of the roots being $5-\sqrt{-1}$.

4. Solve the equation
$$x^5 - 4x^3 + 4x^2 + 4x - 8 = 0,$$
one of the roots being $\sqrt{2}$, and another $1+\sqrt{-1}$.

5. What can be affirmed of the roots of equations under the following conditions:

(1) Coefficients all positive;

(2) Coefficients of the even powers of x preceded by the same sign, and the coefficients of the odd powers preceded by the contrary sign;

(3) Equation containing only the even powers of x, and the coefficients having the same sign;

(4) Equation containing only the odd power of x, and the coefficients having the same sign.

NOTE. The absolute term is regarded as an even power term.

150. Rule of signs. In any series of quantities given with their signs a succession of two like signs is called a **permanence** of signs, and a succession of two unlike signs, a **variation**. Thus, if we write in order the signs of the terms of the polynomial $x^3 - x^2 - 1$, we have $+ - -$; the first and second constitute a variation, the

THEORY OF EQUATIONS. 151

second and third a permanence. The following is known as Descartes' **rule of signs** for positive roots:

No equation can have more positive roots than it has variations in the terms of its first member.

Suppose that the signs of the terms of the polynomial $f(x)$ are

$$+ - + - - - + + -.$$

The signs of a binomial of the form $x - a$, where a is a positive root, are $+ -$. We proceed to show that if the polynomial be multiplied by the binomial, there will be at least one more variation in the product than in the original polynomial.

Writing down the signs only of the terms in the multiplication, and placing an ambiguous sign wherever two terms with different signs are to be added, we have

$$
\begin{array}{c}
+ - + - - - + + - \\
+ - \\
\hline
+ - + - - - + + - \\
 - + - + + + - - + \\
\hline
+ - + - \pm \pm + \pm - +
\end{array}
$$

In this product, we notice that

(1) an ambiguity occurs whenever $+$ follows $+$, or $-$ follows $-$, in the original polynomial;

(2) the signs before and after an ambiguity, or set of ambiguities, are unlike;

(3) a change of sign is introduced at the end;

(4) in general, the sum of the permanences and variations is $n + 1$ if the sum was n in the original polynomial.

On account of (2), the number of variations is not diminished, and in whatever way this series of signs is read, the number of variations is at least one greater than in the original series; and this is true even if we omit the ambiguous signs altogether in reading, the variation being gained at the end.

If the original series ends with a permanence, it will be seen, as before, that there is an additional variation. Thus we may have

$$\begin{array}{c} +\ -\ +\ -\ -\ -\ +\ + \\ +\ - \\ \hline +\ -\ +\ -\ -\ -\ +\ + \\ -\ +\ -\ +\ +\ +\ -\ - \\ \hline +\ -\ +\ -\ \pm\ \pm\ +\ \pm\ - \end{array}$$

Therefore, if we suppose the binomial factors formed from the negative and imaginary roots to have been already multiplied together so that $\phi(x)$ contains only these roots, each factor of the form $x - a$ corresponding to a positive root, introduces at least one variation.

151. If, in the expression

$$f(x) = (x - a_1)(x - a_2) \cdots (x - a_n),$$

we write $-x$ for x, we shall have

$$f(-x) = (-x - a_1)(-x - a_2) \cdots (-x - a_n)$$
$$= (-1)^n (x + a_1)(x + a_2)(x + a_3) \cdots (x + a_n).$$

Whence it appears that $-a_1, -a_2, \cdots -a_n$ are the roots of the equation $f(-x) = 0$, and therefore the roots of $f(-x) = 0$ are equal to those of $f(-x) = 0$ with the signs changed; *i.e.* the positive roots of $f(-x) = 0$ are the negative roots of $f(x) = 0$.

But by the first rule, $f(-x) = 0$ cannot have more positive roots than there are variations in the signs of the terms of $f(-x)$; which is the same as saying that the equation $f(x) = 0$ *cannot have more negative roots than there are variations in the signs of* $f(-x)$. This is Descartes' rule of signs for negative roots.

The student is cautioned against drawing the conclusion that an equation will necessarily have *as many* positive roots as there are variations in the signs of $f(x)$, and as many negative roots as there are variations in the signs of $f(-x)$. It will be observed that, in general, Descartes' rule merely furnishes us with *outside limits* of the number of positive and negative roots; other considerations must determine the number of roots of each kind as well as their value.

EXAMPLES.

1. Find the superior limit to the number of real roots of the equation
$$x^3 + 3x + a = 0,$$
(1) when a is positive, and (2) when a is negative.

2. Find the superior limit to the number of real roots of the equation
$$x^8 - ax^5 + bx - c = 0,$$
a, b, and c being essentially positive.

[The equations of examples 1 and 2 are of great importance in astronomy]

3. Show that the equation
$$x^7 - x^4 + x^3 - 1 = 0$$
has at least four imaginary roots.

4. Find the least possible number of imaginary roots of the equation
$$x^9 - x^5 + x^4 + x^2 + 2 = 0.$$

5. Examine the equation
$$x^n - 1 = 0$$
for real and imaginary roots (1) when n is even, and (2) when n is odd.

6. Examine $\quad x^n + 1 = 0$,
(1) when n is even, and (2) when n is odd.

7. Given $\quad x^3 \pm ax + b = 0$,
where a and b are essentially positive; show that in the first equation we have one negative and two imaginary roots; and in the second, one negative root, while the other two are both imaginary or both positive.

152. Whenever it is required to find the value of $f(x)$ for some particular value of x, the result may be obtained more readily than by direct substitution. Thus if
$$x^4 + p_1x^3 + p_2x^2 + p_3x + p_4$$
is to be evaluated for $x = \alpha$, the process is as follows: write the detached coefficients with their signs, multiply 1, the coefficient of the highest power of x, by α, and add the result to the next coefficient; multiply this sum by α, add the product to the third coefficient, and so proceed till the last term has been used. The last sum is the value of $f(x)$ when $x = \alpha$. The reasons for this rule will be found in considering the process itself as exhibited below.

$$
\begin{array}{cccc}
1 & +p_1 & +p_2 & +p_3 \\
& \alpha & \alpha^2 + p_1\alpha & \alpha^3 + p_1\alpha^2 + p_2\alpha \\
\hline
& \alpha + p_1 & \alpha^2 + p_1\alpha + p_2 & \alpha^3 + p_1\alpha^2 + p_2\alpha + p_3 \\
\end{array}
$$

$$
\begin{array}{c}
+ p_4 \\
\alpha^4 + p_1\alpha^3 + p_2\alpha^2 + p_3\alpha \\
\hline
\alpha^4 + p_1\alpha^3 + p_2\alpha^2 + p_3\alpha + p_4
\end{array}
$$

THEORY OF EQUATIONS. 155

It is evident, also, that if any given function of x is incomplete, *i.e.* if any power of x is absent, its coefficient 0 must be written in the series of coefficients.

Ex. 1. Find the value of $x^5 - 2x^3 + 3x + 4$ when $x = 2$.

1	0	-2	0	$+3$	$+4$	(2
	2	4	4	8	22	
2	2	4	11	26		

$$\therefore f(2) = 26.$$

Ex. 2. Determine whether -3 is a root of
$$x^4 + 7x^3 + 5x^2 - 31x - 30 = 0.$$

1	$+7$	$+5$	-31	-30	$(-3$
	-3	-12	21	30	
1	4	-7	-10	0	

Since $f(-3) = 0$, -3 is a root of $f(x) = 0$.

153. The process of the preceding article also enables us to divide $f(x)$ by $x - \alpha$ when α is a root of $f(x) = 0$.

Let $f(x)$ be evaluated for α as before; the required quotient $f(x) \div (x - \alpha)$ will be, supposing $f(x)$ is of the fourth degree,

$$x^3 + (\alpha + p_1)x^2 + (\alpha^2 + p_1\alpha + p_2)x + (\alpha^3 + p_1\alpha^2 + p_2\alpha + p_3).$$

The function is one degree lower than $f(x)$; and the coefficients of x after the first are the successive sums obtained in the process of evaluating, the last sum vanishing, since α is now supposed to be a root of $f(x) = 0$.

That this is the quotient when $f(x)$ is divided by $x - \alpha$ appears at once by multiplying this new function by $x - \alpha$; thus:

$$\frac{x^3+(\alpha+p_1)x^2+(\alpha^2+p_1\alpha+p_2)x+(\alpha^3+p_1\alpha^2+p_2\alpha+p_3)}{x-\alpha}$$

$$x^4+(\alpha+p_1)x^3+(\alpha^2+p_1\alpha+p_2)x^2+(\alpha^3+p_1\alpha^2+p_2\alpha+p_3)x$$
$$-\alpha x^3-(\alpha^2+p_1\alpha)x^2-(\alpha^3+p_1\alpha^2+p_2\alpha)x-(\alpha^4+p_1\alpha^3+p_2\alpha^2+p_3\alpha)$$

$$x^4+p_1x^3+p_2x^2+p_3x-(\alpha^4+p_1\alpha^3+p_2\alpha^2+p_3\alpha)$$

Since $\quad \alpha^4+p_1\alpha^3+p_2\alpha^2+p_3\alpha+p_4=0,$

or $\quad\quad \alpha^4+p_1\alpha^3+p_2\alpha^2+p_3\alpha=-p_4,$

we have finally for the product

$$x^4+p_1x^3+p_2x^2+p_3x+p_4$$

the original function.

To illustrate the use of this principle let it be required to obtain an equation of lower degree than

$$x^4+7x^3+5x^2-31x-30=0.$$

Evaluating for -3 as above, we may write at once

$$x^3+4x^2-7x-10,$$

the quotient that would be obtained by dividing

$$x^4+7x^3+5x^2-31x-30$$

by $x+3$ in the usual way.

CHAPTER XX.

GRAPHS.

154. A root of $f(x) = 0$ has already been defined as a value of x which will render the statement, $f(x) = 0$, true; that is, if α be a root, $f(\alpha)$ vanishes. If any other value not a root, as β, be assigned to x, the function thus evaluated will not vanish; *i.e.* we shall have $f(\beta)$ equal to some quantity not zero.

Now if x be viewed as a variable and be conceived to pass through all real values from $-\infty$ to $+\infty$, it is evident (1) that we shall have as many values for the function as for the variable, (2) that in general the value of the function will not be zero, and (3) that whenever x reaches a *root* value the function will be zero.

It is possible to represent graphically the changes in the function corresponding to the changes in the variable. To do it, draw a horizontal line with a second straight line at right angles to the first. Call the horizontal line, XX', or the X-axis; the vertical line YY', or the Y-axis; and their point of intersection O, or the origin.

Beginning at O as the zero point, lay off with any convenient unit of length the positive values of x to the right on the X-axis; and the negative values to the left on this axis. At the end, remote from O, of this line which represents the value of x, draw a perpendicular (using the same unit of length) to represent the corresponding value of $f(x)$. The perpendicular is to be

158   ALGEBRA.

drawn upward from the X-axis in case $f(x)$ is positive, and downward from this axis when $f(x)$ is negative.

Thus a point is located in the plane, and if many values be given to x, — any two consecutive values differing but little from each other, — we shall have a correspondingly large number of points with small distances separating them; and the assemblage of points will ultimately form a continuous line, straight or curved, which may be called the **graph** of $f(x)$. The values of x are called **abscissas**, and the values of $f(x)$, **ordinates**; the two together are known as **co-ordinates**.

To illustrate: let us take the equation $x + 2 = 0$ and construct the graph of $f(x)$, having in this case $f(x) = x + 2$. When $x = 0$, $f(x) = +2$; when $x = +1$,

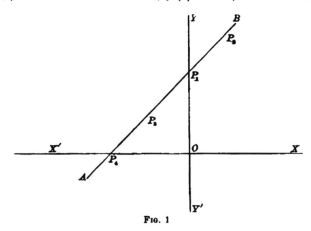

Fig. 1

$f(x) = +3$; when $x = +2$, $f(x) = +4$; etc. Taking the first pair of values, since $x = 0$ we have no distance to measure off on the X-axis, and since $f(x) = +2$, we measure upward two units, thus locating the graphic point P_1. Again measuring one unit to the right and three units upward, we have the point P_2. If $x = -1$, $f(x) = +1$; hence, measuring one unit to the left on the X-axis and then one unit upward, we have a third point P_3.

If many points are located in this way, they will be found to lie on the straight line determined by any two of them. (It will be shown in analytic geometry that the graph is always a straight line when $f(x)$ is of the first degree.) We shall therefore draw the straight line AB passing through the points determined.

For our present purposes, especial attention must be given to the fact that *the distance from the origin O to the point where the graph cuts the X-axis represents a root of* $f(x) = 0$.

This is evidently true, since only root values of x make $f(x)$ vanish, and when $f(x)$ vanishes, we have to measure neither up nor down from the X-axis; that is, the points determined are the particular points of the graph which lie on the X-axis.

In the above example the distance $OP_4 (= -2)$ represents the one root of the equation $x + 2 = 0$.

It will now be seen that having any equation $f(x) = 0$, of any degree, if we can construct its graph, the several distances from O to the points where the graph cuts the X-axis will be the real roots of the equation; we are thus furnished with one method of solution of equations.

Ex. 2. Construct the graph of $f(x)$ when
$$f(x) = x^2 + x - 6;$$
and by means of it show that the roots of the equation $x^2 + x - 6 = 0$ are -3 and $+2$.

Writing a value of x and the accompanying value of $f(x)$ in parentheses, we have, for instance, the following pairs of values: $(0, -6)$; $(1, -4)$; $(-1, -6)$; $(-3, 0)$; $(2, 0)$.

160   *ALGEBRA.*

Locating the corresponding points as in the first example, and drawing a curve through them, the graph is as seen in Fig. 2. If many points be located, they will be found to lie on the curve as drawn; there are no small bends or other irregularities in the curve. Moreover, the graph of any particular quadratic function, when determined in this experimental manner, will be found to be similar to the graph in Fig. 2.

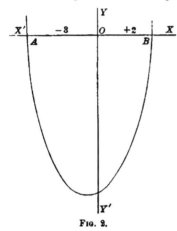

Fig. 2.

In analytic geometry it is proven deductively that the graphs of *all* functions of the form $ax^2 + bx + c$ are similar to that of Fig. 2.

155. If we move any graph upwards through one unit, or what would amount to the same thing, move the X-axis down through one unit keeping it parallel to itself, we increase every value of the function by one unit. But we may increase the value of the function by one unit by adding a unit to the absolute term.

Hence, the geometrical signification of the algebraic operation of increasing the absolute term of $f(x)$ by any quantity as p_{n+1} is this: the graph of $f(x)$ is moved upwards through the distance p_{n+1}; and if the absolute term is diminished by p_{n+1}, the graph is moved downwards through the distance p_{n+1}.

Now if any graph, such as the one in Fig. 2, is moved upwards the two quantities which OA and OB represent

GRAPHS. 161

are approaching equality; and if we move the graph upwards just far enough it will merely touch the line XX' and the points A and B are infinitely near to each other. Any further motion upwards brings the graph to a position so that the line XX' does not touch it at all, and we say that the X-axis cuts the graph in two imaginary points.

Since this motion of the graph was accomplished by increasing the absolute term, and since the intercepts OA and OB constantly represent the roots of the equation $f(x) = 0$, it follows that an increase in the absolute term causes a pair of unequal real roots to change so that they first become a pair of equal real roots, and finally become conjugate imaginary roots.

If the graph had another bend in it, a bend above the X-axis and convex upward so that the graph should cross the X-axis at C to the left of A, we should need to move the whole graph downwards in order to change the two unequal real roots, OA and OC, into two equal real roots and then into conjugate imaginaries.

Thus in Fig. 3, if the graph marked $f(x)$ is moved downwards, the intercepts OP and OP' approach equality; that is, two roots of the equation whose absolute term is being diminished are approaching the limit of their real values, and when the bend $PB'B''P'$ is below the X-axis and free from it, the roots in question are two conjugate imaginaries.

To illustrate this change in roots, if $\frac{25}{4}$ be added to the absolute term of the equation in the above example, we have $x^2 + x + \frac{1}{4} = 0$, an equation whose roots are $-\frac{1}{2}$ and $-\frac{1}{2}$; the two unequal intercepts, -3 and $+2$, on the X-axis have been replaced by the two coincident intercepts $-\frac{1}{2}$ and $-\frac{1}{2}$.

If a still greater change be made in the absolute term

of $x^2 + x - 6$, the graph will no longer touch the X-axis; the two equal roots of the equation will be replaced by two imaginary roots, and the graph is cut by the X-axis in two imaginary points. For instance, if the absolute term $\frac{1}{4}$ were changed to $\frac{1}{2}$, we should have the equation $x^2 + x + \frac{1}{2} = 0$, whose roots are $-\frac{1}{2} \pm \frac{1}{2}\sqrt{-1}$.

156. "The mode of representing a function by a graph is due to Descartes, and its invention is one of the great milestones in the progress of mathematics. The graph is largely employed by statisticians, by engineers, by physicists, by chemists, and by many others who are able to employ mathematical methods intelligently; and its systematic discussion is the subject-matter of co-ordinate (analytic) geometry."*

EXAMPLES.

1. Construct the graph of $x^2 - 4x + 4$, and show that the equation
$$x^2 - 4x + 4 = 0$$
has two equal roots.

2. Construct the graph of $x^2 - 2x + 4$, and show that the equation
$$x^2 - 2x + 4 = 0$$
has no real roots.

Compare examples 3 and 4, and notice that a change in the position of the graph with respect to the X-axis may be accomplished in other ways besides a change in the absolute term.

3. Construct the graph of $x^3 - 6x^2 + 11x - 6$, and by means of it show the roots of the equation
$$x^3 - 6x^2 + 11x - 6.$$

* N. F Dupuis' *Principles of Elementary Algebra*.

4. Show that the equation
$$x^3 + 2x^2 + 4x + 8 = 0$$
has only one real root, and that this root is negative.

5. Construct the graph of
$$x^4 - 6x^3 + 13x^2 - 12x + 4,$$
and show that the equation formed by equating this function to zero has two pairs of equal roots.

6. Form an equation of the fourth degree in real coefficients that shall have two of its roots zero and a third root $\frac{1}{2}\sqrt{-1}$; and exhibit the roots graphically.

CHAPTER XXI.

SPECIAL ROOTS.

157. In this chapter it is proposed to examine conditions whose conclusions relate to real roots, equal roots, roots which are multiples of other roots, and roots which are infinitely great.

158. Suppose that
$$\alpha - h, \ \alpha, \ \alpha + h$$
are three values of x taken in order, and that α is a root of the equation $f(x) = 0$.

Then the three accompanying values of the function are
$$f(\alpha - h), \ f(\alpha), \ f(\alpha + h),$$
of which $f(\alpha)$ is zero and the other two are not zero.

It is required to compare the signs of $f(\alpha - h)$ and $f(\alpha + h)$,

(1) when there is an *odd* number of roots having the value α;

(2) when there is an *even* number of roots having this value.

To make the case perfectly general, let $\alpha_1, \alpha_2 \cdots \alpha_{n-2m}$ be the real roots, and let $\beta_1, \beta_2, \cdots \beta_{2m}$ be the imaginary roots, and suppose
$$\phi(x) = (x - \beta_1)(x - \beta_2) \cdots (x - \beta_{2m}).$$

SPECIAL ROOTS.

Then we have
$$f(x) = (x - \alpha_1)(x - \alpha_2) \cdots (x - \alpha_{n-2m}) \{\phi(x)\} = 0.$$

As $f(x)$ is supposed to contain only real coefficients, $\phi(x)$ must be of even degree. No real value of x can cause $\phi(x)$ to change sign; for if we form the equation $\phi(x) = 0$, it has only imaginary roots; therefore the graph of $\phi(x)$ does not cut the X-axis, and all the perpendiculars representing values of $\phi(x)$ are on the same side of the axis. Hence, in examining $f(x)$ for changes of sign, we need consider only the product
$$(x - \alpha_1)(x - \alpha_2) \cdots (x - \alpha_{n-2m}).$$

Suppose x to start with some value less than the least root, and to increase continuously until it becomes greater than the greatest root. So long as x is less than the least root, all the factors $(x - \alpha_1)$, $(x - \alpha_2)$, etc., are negative; but when x passes the value of the least root, say α_1, the factor containing that root becomes positive, and if there is no other root equal to α_1, this factor will be the only one which will change sign; consequently the sign of the entire product is changed.

Further, if there is an odd number of roots having the value α_1, an *odd* number of factors will change sign, and the sign of the entire product is changed; but if there is an *even* number of roots having the value α_1, an even number of factors will change sign, and therefore the sign of the entire product remains unchanged.

As x goes on increasing after it has passed the first root, it is evident that the factor containing this root cannot become negative again, and does not need to be considered in determining the subsequent signs of $f(x)$.

The above argument is repeated in the case of each root $\alpha_2, \alpha_3, \cdots \alpha_{n-2m}$, and thus the theorem is established that f(x), *of the equation* f(x) = 0, *will change sign when* x *passes through any real root, as* a, *if there is an odd number of roots having the value* a; *but if there is an even number of such roots,* f(x) *will not change sign.*

159. Suppose that upon evaluating $f(x)$ in succession for two quantities, as a and b, the results $f(a)$ and $f(b)$ have unlike signs: we have then, on the graph of $f(x)$, two points whose co-ordinates are $[a, f(a)]$ and $[b, f(b)]$, and since it is assumed that $f(a)$ and $f(b)$ have unlike signs, the two points must lie on opposite sides of the X-axis. Now $f(x)$ is of the form

$$x^n + p_1 x^{n-1} + \cdots + p_{n-1} x + p_n,$$

a rational integral algebraic function, and no finite value of x can render the function equal to ∞. Moreover, as shown in Art. 133, $f(x)$ varies continuously from $f(a)$ to $f(b)$; *i.e.* it passes through all intermediate values while x changes from a to b; hence the graph must cross the X-axis in order to connect the points $[a, f(a)]$ and $[b, f(b)]$, but at the point of crossing $f(x) = 0$, and there is therefore a root of $f(x) = 0$ between a and b.

It evidently follows that the graph, by crossing the X-axis an *odd* number of times, may connect points lying on opposite sides of the axis. Therefore if $f(a)$ and $f(b)$ have unlike signs, there may be more than one real root of $f(x) = 0$ between the values a and b but the number of them is an odd number.

We thus conclude that *if* f(x) *be evaluated in succession for two quantities,* a *and* b, *and if the results,* f(a) *and* f(b), *have unlike signs, the equation* f(x) = 0 *must have at least one real root between* a *and* b.

For example,

if $f(x) = x^3 + x - 1,$

$f(-\tfrac{1}{2}) = -\tfrac{13}{8},$ and $f(1) = 1;$

hence there is a root of the equation
$$x^3 + x - 1 = 0$$
between $-\tfrac{1}{2}$ and 1; and the graph of $x^3 + x - 1$ must cross the X-axis at least once between the point $x = -\tfrac{1}{2}$ and the point $x = 1$.

160. It has been shown in a former article that an equation whose coefficients are all real may have pairs of conjugate imaginary roots. In case $f(x)$ is wholly made up of the product of quadratic functions of the form $(x - \alpha)^2 + \beta^2$, there is then no real quantity which substituted for x makes $f(x)$ vanish; and since $f(x) = 0$ has thus no real roots, the graph of $f(x)$ must lie altogether on one side of the X-axis; and if we can determine the sign of $f(x)$ for one value of x, we shall know its sign for all values of x. The equation $f(x) = 0$, where
$$f(x) = x^n + p_1 x^{n-1} + \cdots + p_{n-1} x + p_n,$$
may be written
$$x^n \left\{ 1 + p_1 \frac{1}{x} + p_2 \frac{1}{x^2} + \cdots + p_{n-1} \frac{1}{x^{n-1}} + p_n \frac{1}{x^n} \right\} = 0.$$

As x increases, the value of the function tends to become x^n, and when $x = +\infty$, $f(x) = (+\infty)^n$, a positive quantity infinitely great.

Hence, *if there exists no real quantity which, substituted for* x, *makes* f(x) *vanish, then* f(x) *must be positive for every real value of* x.

For example, the graph of the fourth degree function
$$[(x - 1)^2 + 2][(x + 2)^2 + 3]$$
must lie altogether above the X-axis.

161. Under the condition stated in the last article, the function is necessarily of even degree, being composed of quadratic functions. Suppose now that

$$f(x) = x^n + p_1 x^{n-1} + \cdots + p_{n-1} x + p_n = 0,$$

n being an odd number. Since imaginary roots enter the equation in pairs, if $2m$ be the number of imaginary roots, the number of real roots is $n - 2m$, an odd number. It can now be shown that one of these $n - 2m$ real roots is of opposite sign to that of p_n.

If $\begin{cases} x = -\infty, \ f(x) \text{ is negative;} \\ x = 0, \text{ the sign of } f(x) \text{ is the same as that of } p_n; \\ x = +\infty, \ f(x) \text{ is positive.} \end{cases}$

If p_n is positive, the graph of $f(x)$ must cross the X-axis between $-\infty$ and 0; that is, there is one negative root; and if p_n is negative, the graph crosses the axis between 0 and $+\infty$; that is, there is a root between 0 and $+\infty$.

Hence, *every equation of an odd degree has at least one real root of a sign opposite to that of its last term.*

162. Let it now be assumed that an equation is of an even degree and that its last term is negative.

Substituting as in the preceding demonstration, we have:

x	$f(x)$
$-\infty$	$+$
0	$-$
$+\infty$	$+$

SPECIAL ROOTS. 169

Since $f(x)$ has changed sign between $-\infty$ and 0, and again between 0 and $+\infty$, there must exist at least one real negative root and one real positive root.

Hence, *every equation of an even degree whose last term is negative has at least two real roots, one positive and one negative.*

163. As a preliminary to a method for testing an equation for equal roots, we shall show that if the equation $f(x) = 0$ has no equal roots, $f(x)$ and $f'(x)$ have no H. C. F. (see Art. 70).

Let $f(x) = (x - \alpha_1)(x - \alpha_2) \cdots (x - \alpha_n)$,

in which no two of the quantities $\alpha_1, \alpha_2 \cdots \alpha_n$ are equal to each other.

Then $f'(x) = +(x - \alpha_2)(x - \alpha_3) \cdots (x - \alpha_n)$
$+ (x - \alpha_1)(x - \alpha_3) \cdots (x - \alpha_n) + \cdots$
$+ (x - \alpha_1)(x - \alpha_2) \cdots (x - \alpha_{n-1})$.

No one of the linear binomial factors of $f(x)$ is a factor of $f'(x)$; and since these linear factors are the prime factors, the two functions $f(x)$ and $f'(x)$ have no H. C. F.

164. Equal roots. If we make a supposition the opposite of that made in the preceding article, we find that $f(x)$ and $f'(x)$ have an H. C. F.

For let α^1 be one of the m equal roots of $f(x) = 0$, and let the other roots be $\alpha_2, \alpha_3, \cdots \alpha_{n-m+1}$; then

$$f(x) = (x - \alpha_1)^m (x - \alpha_2) \cdots (x - \alpha_{n-m+1})$$
$$= (x - \alpha_1)^m \phi(x).$$

Hence $f'(x) = m(x - \alpha_1)^{m-1} \phi(x) + (x - \alpha_1)^m \phi'(x)$.

By the preceding article, $\phi(x)$ and $\phi'(x)$ have no H.C.F.; therefore $(x - a_1)^{m-1}$ is the H. C. F. of $f(x)$ and $f'(x)$.

Now if we put this H. C. F. equal to zero, we have

$$(x - a_1)^{m-1} = 0,$$

an equation of the $(m-1)$th degree, having for its roots a_1 taken $m-1$ times. Thus the number of equal roots in the given equation is greater by one than the number obtained from the equation $(x - a_1)^{m-1} = 0$.

Hence, *if* f(x) $= 0$ *has equal roots, the H. C. F. of* f(x) *and* f'(x), *when equated to zero, constitutes an equation which has for its roots these equal roots, and no others.*

If $f(x) = 0$ has two sets of equal roots, so that

$$f(x) = (x - a_1)^m (x - a_2)^q \phi(x),$$

the above process may be repeated, and the H. C. F. of $f(x)$ and $f'(x)$ will be found to be

$$(x - a_1)^{m-1}(x - a_2)^{q-1}.$$

The solution of the equation,

$$(x - a_1)^{m-1}(x - a_2)^{q-1} = 0,$$

will evidently give roots having the values a_1 and a_2, $m-1$ roots with the former value and $q-1$ with the latter value.

The process will be the same for any number of sets of equal roots.

165. If $f(x) = 0$ has m equal roots, $f'(x) = 0$ has $m-1$ equal roots whose value is the same as that of one of the equal roots of the first equation.

If an equation is suspected of having equal roots, the H. C. F. of $f(x)$ and $f'(x)$ should be found.

Suppose, for example, that the factor is $x^2 - 4x + 4$;

SPECIAL ROOTS. 171

the roots of $x^2 - 4x + 4 = 0$ being 2 and 2, we have now $m - 1$ of the m equal roots which the given equation $f(x) = 0$ must have; hence 2, 2, 2 are roots of the equation, and since $f(x)$ is divisible by $(x - 2)(x - 2)(x - 2)$, we may, by dividing, reduce the degree of the equation so that there is left for solution an equation whose degree is $(n - 3)$.

166. Thus far we have considered only such equations as have unity for the coefficient of the highest power of x. In case the coefficient of the highest power is not unity, the equation may be transformed into one in which the coefficient of the highest power is unity and the other coefficients are integral, the original coefficients being integral.

Suppose we have

$$p_0 x^n + p_1 x^{n-1} + p_2 x^{n-2} + \cdots + p_{n-1} x + p_n = 0;$$

then $\quad x^n + \dfrac{p_1}{p_0} x^{n-1} + \dfrac{p_2}{p_0} x^{n-2} + \cdots + \dfrac{p_{n-1}}{p_0} x + \dfrac{p_n}{p_0} = 0.$

Now let $x = \dfrac{y}{A}$, in which A is at first an undetermined quantity.

Substituting this value of x,

$$\left(\frac{y}{A}\right)^n + \frac{p_1}{p_0}\left(\frac{y}{A}\right)^{n-1} + \cdots + \frac{p_{n-1}}{p_0}\left(\frac{y}{A}\right) + \frac{p_n}{p_0} = 0.$$

Multiplying by A^n,

$$y^n + \frac{A p_1}{p_0} y^{n-1} + \cdots + \frac{A^{n-1} p_{n-1}}{p_0} y + \frac{A^n p_n}{p_0} = 0.$$

If, now, A be taken equal to p_0, we have an equation of the form

$$y^n + P_1 y^{n-1} + P_2 y^{n-1} + \cdots + P_{n-1} y + P_n = 0,$$

in which the coefficients P_1, P_2, etc., are all integral. After solving this equation in y, the root-values of x in the original equation may be found from the relation $x = \dfrac{y}{A}$.

167. Roots infinitely great. It is sometimes necessary to impose upon an equation, whose coefficients are literal or partly literal, the condition that one or more of its roots shall be infinitely great.

In order to do this, let $x = \dfrac{1}{y}$, and substitute this value of x in the equation

$$p_0 x^n + p_1 x^{n-1} + \cdots + p_{n-1} x + p_n = 0; \qquad (1)$$

then we have

$$p_0 \left(\frac{1}{y}\right)^n + p_1 \left(\frac{1}{y}\right)^{n-1} + \cdots + p_{n-1}\left(\frac{1}{y}\right) + p_n = 0,$$

or $\qquad p_0 + p_1 y + \cdots + p_{n-1} y^{n-1} + p_n y^n = 0. \qquad (2)$

If p_0, which has now become the absolute term, be of the nature of an undetermined coefficient, we may equate it to zero, and equation (2) then becomes

$$y(p_1 + p_2 y + \cdots + p_{n-1} y^{n-2} + p_n y^{n-1}) = 0;$$

and consequently, zero is one of its roots. But when $y = 0$, $x = \infty$.

Hence, *the condition that one of the roots of an equation shall be infinity is that the coefficient of the highest power of the unknown quantity shall be zero.* If both p_0 and p_1 are zero, two roots of equation (2) are zero; hence two roots of (1) are infinitely great.

The principle of this article is of importance in the theory of asymptotes. See Williamson's *Differential Calculus*, Chapter XIII.

SPECIAL ROOTS. 173

168. Identical equations. Two equations are said to be identical if their corresponding coefficients are proportional.

Thus $\quad p_0 x^n + p_1 x^{n-1} + \cdots + p_{n-1} x + p_n = 0 \quad (1)$

and $\quad P_0 x^n + P_1 x^{n-1} + \cdots + P_{n-1} x + P_n = 0 \quad (3)$

are identical if

$$\frac{p_0}{P_0} = \frac{p_1}{P_1} = \cdots = \frac{p_{n-1}}{P_{n-1}} = \frac{p_n}{P_n},$$

and equation (3) might have been derived from (1) by multiplying the first term by $\dfrac{P_0}{p_0}$, the second by $\dfrac{P_1}{p_1}$, and so on. See Art. 78.

EXAMPLES.

1. Find the roots of the following equations:
    1. $x^3 - 3x^2 - 6x + 8 = 0$.
    2. $x^4 - 2x^2 + 1 = 0$.
    3. $x^4 + 4x^3 + x^2 - 8x - 6 = 0$
    4. $x^5 + x^4 - 8x^3 - 6x^2 = 0$.

2. Examine the following equations for equal roots.
    1. $x^4 - 4x^3 + 3x^2 + 4x - 4 = 0$.
    2. $x^4 - 6x^3 + 6x^2 + 18x - 27 = 0$.
    3. $x^6 - 6x^5 + 3x^4 + 12x^3 - 9x^2 - 6x + 5 = 0$.
    4. $x^3 - 5x^2 + 3x + 9 = 0$.

3. Find the condition that the equation
$$x^3 + 3ax^2 + 36x + c = 0$$
shall have two roots equal

4. Find the condition that the equation
$$x^4 + ax^2 + b = 0$$
shall have equal roots, and determine whether it can have three equal roots.

5. Show the general features of the graphs for equations of the following description:

1. An equation of the third degree, all of its roots real and two of them equal.

2. An equation formed by equating to zero the first derivative of the function in the preceding example.

3. An equation of the third degree, all of its roots equal and positive.

4. An equation of the fourth degree, all of its roots imaginary.

6. Find the roots of the following equations:
   1. $10x^3 - 17x^2 + x + 6 = 0$.
   2. $3x^3 - 2x^2 - 6x + 4 = 0$.
   3. $2x^3 - 3x^2 + 2x - 3 = 0$.
   4. $x^4 - \frac{1}{2}x + \frac{3}{16} = 0$.

7. Given the equations:
   1. $(mx + n)(x^2 + 3a^2) = x^3$.
   2. $(mx + n)^3 - (6x^2 - x^3) = 0$.
   3. $(mx + n)^3 + x^3 - a^3 = 0$.
   4. $x^3 + (mx + n)^3 = 3ax(mx + n)$.

Perform the indicated operations and arrange the terms in descending powers of x; then impose the condition that each equation shall have two roots infinitely great, and find the resulting values of m and n in each case.

CHAPTER XXII.

STURM'S THEOREM.

169. Let
$$f(x) = (x - \alpha_1)(x - \alpha_2)(x - \alpha_3) \cdots (x - \alpha_n) = 0.$$

If the multiplication here indicated be performed, we have

$$x^n - (\alpha_1 + \alpha_2 + \cdots + \alpha_n)x^{n-1} + (\alpha_1\alpha_2 + \alpha_1\alpha_3 + \alpha_2\alpha_3 + \cdots)x^{n-2}$$
$$+ \cdots + (-1)^n(\alpha_1\alpha_2\alpha_3 \cdots \alpha_n) = 0.$$

It will be seen that the coefficient of x^n is $+1$; the coefficient of x^{n-1} is the sum of the roots, preceded by the minus sign; the coefficient of x^{n-2} is the sum of the product of the roots taken two and two; the coefficient of x^{n-3} is the sum of the product of the roots taken three and three, and preceded by the minus sign.

The law of the formation of the coefficients is now apparent, and we have finally for the last term the product of all the roots, preceded by the minus sign if there is an odd number of roots. If this equation be identical with

$$x^n + p_1 x^{n-1} + p_2 x^{n-2} + \cdots + p_{n-1} x + p_n = 0,$$

by Art. 168 we have

$$-(\alpha_1 + \alpha_2 + \cdots + \alpha_n) = p_1,$$
$$+(\alpha_1\alpha_2 + \alpha_1\alpha_3 + \alpha_2\alpha_3 + \cdots) = p_2,$$
$$\cdots \cdots \cdots \cdots \cdots \cdots$$
$$(-1)^n(\alpha_1\alpha_2\alpha_3 \cdots \alpha_n) = p_n;$$

176   ALGEBRA.

a set of n independent equations in the n unknown quantities $\alpha_1, \alpha_2, \cdots \alpha_n$, the known quantities being $p_1, p_2, \cdots p_n$.

Now it is obvious from the definition of a root, and in advance of investigation, that a root of an equation must be some function of the coefficients in that equation; and it would seem as if the above set of equations furnished us with material for determining *what* functions the roots are of the coefficients; that is, it would seem as if we might solve these n equations, obtaining the value of each root. A general solution would, of course, furnish formulas by means of which the roots could be determined in any particular case.

Since the quantities $\alpha_1, \alpha_2, \alpha_3$, etc., are all involved in the same way in these n equations, there is no advantage in eliminating certain quantities rather than others.

Suppose the $n-1$ quantities $\alpha_2, \alpha_3, \alpha_4, \cdots \alpha_n$, eliminated so that we have finally an equation in α_1. In whatever way the elimination is effected, this equation is of the nth degree in α_1; in fact, it presents the original function with α_1 in place of x; hence no progress is made toward the solution of the given equation in x.

170. The principles of previous chapters are of service in the determination of the roots of equations, but it has been seen that some of the principles apply only to equations having special characteristics, while others afford methods which are more or less tentative. We proceed to a theorem which enables us to determine the number and situation of the real roots of any numerical equation of the form

$$x^n + p_1 x^{n-1} + \cdots + p_{n-1} x + p_n = 0,$$

n being positive and integral, and the equation having no

equal roots. This theorem, named from its discoverer, Sturm,* is as follows:

Given the equation $f(x) = 0$ *without equal roots. Let the operation of finding the H. C. F. of* $f(x)$ *and* $f'(x)$ *be performed with this modification, that the sign of every remainder is changed before it is used as a divisor. The given equation having no equal roots, we arrive at last at a numerical remainder not zero. Let the sign of this remainder also be changed. Let* $f_1(x), f_2(x), \cdots f_{n-1}(x)$ *be the series of the modified remainders so that we have for the entire series* $f(x), f'(x), f_1(x), f_2(x), \cdots f_{n-1}(x)$, *the series decreasing in degree from the nth to the zero degree inclusive. Let* a *and* b (*where* b $>$ a) *be any two real quantities; then the number of real roots of* $f(x) = 0$ *between* a *and* b *is equal to the excess of the number of variations in the series of signs of* $f(a), f'(a), f_1(a), \cdots f_{n-1}(a)$ *over the number of variations in the series of signs of* $f(b), f'(b), f_1(b), \cdots f_{n-1}(b)$.

$f_1(x), f_2(x), \cdots f_{n-1}(x)$ may be called **Sturmian functions**. Let $q_0, q_1, \cdots q_{n-2}$ denote the successive quotients which arise in the process of producing the Sturmian functions; then we have

$$f(x) = q_0 f'(x) - f_1(x),$$
$$f'(x) = q_1 f_1(x) - f_2(x),$$
$$f_1(x) = q_2 f_2(x) - f_3(x),$$
$$f_2(x) = q_3 f_3(x) - f_4(x),$$
$$\cdots \cdots \cdots \cdots$$
$$f_{n-3}(x) = q_{n-2} f_{n-2}(x) - f_{n-1}(x).$$

* Ce grand géomètre communiqua à l'Académie des Sciences, en 1829, la démonstration de son théorème qui constitue l'une des plus brillantes découvertes dont se soit enrichie l'Analyse mathématique. — *Serret.*

From these relations two conclusions can be drawn:

First, *two consecutive functions cannot vanish for the same value of* x. Suppose the contrary to be true, and assume that $f_1(x)$ and $f_2(x)$ become equal to zero for the same value of x; then from the third of the above equations, $f_3(x)$ must also equal zero; and if $f_2(x)$ and $f_3(x)$ vanish simultaneously, $f_4(x)$ must also vanish in order that the fourth equation may be true. Finally we shall have $f_{n-1}(x)$ vanishing with the rest; but this last Sturmian function, being a constant, cannot vanish for any value of x.

Second, *when any intermediate function vanishes, the function preceding and the one succeeding have unlike signs.* Suppose, for instance, that $f_3(x)$ vanishes for some value of x, as c. We have just seen that neither $f_2(x)$ nor $f_4(x)$ can vanish simultaneously with $f_3(x)$; hence the fourth equation of the above set becomes $f_2(x) = -f_4(x)$.

Now no alteration can be made in the series of signs of the functions

$$f(x),\ f'(x),\ f_1(x),\ \cdots f_{n-1}(x),$$

except when x passes through a value which makes some function vanish. We proceed to the proof of the following sub-propositions:

1. *No variation is lost or gained in consequence of* x *passing through a value which makes any function except* f(x) *vanish.*

2. *When* x *passes through a value that makes* f(x) *vanish, one variation, and only one, is lost.*

Let c be a value of x which makes some other function than $f(x)$ vanish. For example, suppose $f_3(c) = 0$. As

STURM'S THEOREM.

neither $f_2(x)$ nor $f_4(x)$ vanishes when $x = c$, neither of them can change sign as x passes through c, but $f_3(x)$ will or will not change sign, according as the equation $f_3(x) = 0$ has an odd or even number of equal roots of the value c; hence the signs of $f_3(x)$ just before $x = c$ and just after $x = c$ may be any one of the four cases, $++$, $--$, $+-$, $-+$; hence if $c - h$ be the value of x just before $x = c$, and $c + h$ its value just after $x = c$, and if $f_2(x)$, $f_3(x)$, $f_4(x)$ be evaluated first for $c - h$, and then for $c + h$, all the possible arrangements of signs that can occur are:

$x = c - h$				$x = c + h$		
$f_2(c-h)$	$f_3(c-h)$	$f_4(c-h)$		$f_2(c+h)$	$f_3(c+h)$	$f_4(c+h)$
+	+	−		+	+	−
+	−	−		+	−	−
+	+	−		+	−	−
+	−	−		+	+	−
−	+	+		−	+	+
−	−	+		−	−	+
−	+	+		−	−	+
−	−	+		−	+	+

These signs should be read horizontally; it will be noticed that in every case there is one variation and one permanence.

If $h = 0$, so that $f_3(c \mp h)$ is actually zero, the middle column of signs drops out, and we have simply $+-$ or $-+$; *i.e.* one variation as before. Further, non-consecutive functions may vanish for the same value of x. When x reaches such a value, the number of signs in the series to be considered is at that instant diminished by the number of functions which vanish; but from the above argument it is seen that only *permanences* disappear, and that these re-appear as soon as x has passed that value which causes the several non-consecutive functions to vanish.

We have finally to consider the effect of a passage of x through a root of $f(x) = 0$.

Suppose α a root, so that $f(\alpha) = 0$. Let h be a positive quantity, so that $\alpha - h$ is the value of x just before the assumed root value is reached, and $\alpha + h$ is the value of x just after passing that value. By Art. 132,

$$f(\alpha - h) = f(\alpha) - hf'(\alpha) + \frac{h^2}{\lfloor 2}f''(\alpha) - \frac{h^3}{\lfloor 3}f'''(\alpha) + \cdots;$$

also

$$f(\alpha + h) = f(\alpha) + hf'(\alpha) + \frac{h^2}{\lfloor 2}f''(\alpha) + \frac{h^3}{\lfloor 3}f'''(\alpha) + \cdots.$$

Since $f(\alpha) = 0$, these two equations may be written

$$\frac{f(\alpha - h)}{h} = -f'(\alpha) + \frac{h}{\lfloor 2}f''(\alpha) - \cdots,$$

$$\frac{f(\alpha + h)}{h} = f'(\alpha) + \frac{h}{\lfloor 2}f''(\alpha) + \cdots,$$

or if h be taken indefinitely small,

$$\frac{f(\alpha - h)}{h} = -f'(\alpha), \qquad (1)$$

$$\frac{f(\alpha + h)}{h} = +f'(\alpha). \qquad (2)$$

Now since the equation $f(x) = 0$ has no equal roots, and consequently the equations $f(x) = 0$ and $f'(x) = 0$ have no common root, it follows that $f(x)$ changes sign when x passes through the value α, and $f'(x)$ does not change sign; *i.e.* the quantities $f(\alpha - h)$ and $f(\alpha + h)$ have unlike signs, whilst $f'(\alpha - h)$ and $f'(\alpha + h)$ have like signs.

Summarizing the argument, we have

$$f(\alpha - h)\{\pm, \qquad f'(\alpha - h)\{\mp,$$
$$f(\alpha + h)\{\mp, \qquad f'(\alpha + h)\{\mp.$$

Reading the upper signs of the functions in the first horizontal line, and then the upper signs in the next line, and reading the lower signs in the same manner, we see that in each case a variation is succeeded by a permanence.

Therefore if the entire series of functions $f(x), f'(x),$ $f_1(x), f_2(x)$, etc., be evaluated for $\alpha - h$ where α is a root of $f(x) = 0$, and h is indefinitely small, and if the series be then evaluated for $\alpha + h$, the second series of signs must have one less variation than the first series has; and since one variation must thus give way to a permanence each time that x passes through a root of $f(x) = 0$, and since no variation can be lost as x passes through other values, the theorem is established.

171. In order to find the whole number of real roots of an equation $f(x) = 0$, we first make $x = -\infty$ in the series of functions $f(x), f'(x), f_1(x)$, etc., and note the number of variations; we then make $x = +\infty$, and note the number of variations. The excess of the number of variations in the first series over the number in the second series is the whole number of real roots. If x is made equal to 0 in the series, the excess of the number of variations when $x = -\infty$ over the number when $x = 0$ determines how many of these real roots are negative.

It is to be understood that in forming the Sturmian functions, not only must the sign of each remainder be changed before it is used as a divisor, but a negative factor must *not* be introduced at any other point in the operation.

It is left to the student to show why the remainders *with their signs changed*, rather than the ordinary remainders, are taken for the Sturmian functions.

Since $f(x)$ and $f'(x)$ must have unlike signs just before x reaches each real root of the equation $f(x) = 0$, it follows that the real roots of the equation $f'(x) = 0$ are intermediate between those of $f(x) = 0$. Hence if the equation $f(x) = 0$ is of the nth degree, and has m real roots, $f'(x) = 0$, an equation of the $(n-1)$th degree, has at least $m-1$ real roots.

172. The chief features of the demonstration of Sturm's theorem may be illustrated by means of the graphs of $f(x)$, $f'(x)$, $f_1(x)$, etc.

For example, let
$$f(x) = x^3 - 2x^2 - 6x + 4,$$
then $\quad f'(x) = 3x^2 - 4x - 6,$

and the Sturmian functions $f_1(x)$ and $f_2(x)$ are $11x - 6$ and $+882$. In the construction of the graphs in this instance, no use will be made of Sturm's theorem to discover the roots of $f(x) = 0$. Since the equation is one of an odd degree, it has at least one real root; and since its last term is positive, it has a negative root. Evaluating for negative quantities, we find that -2 is a root. Lowering the degree of the equation and solving the resulting quadratic $x^2 - 4x + 2 = 0$, the other roots are found to be real and positive $(2 \pm \sqrt{2})$.

Since we now have the three points where the graph crosses the X-axis, it may be constructed in its essential features, and, in connection with it, the graph of $f'(x)$, the roots of $f'(x) = 0$ being $\frac{2}{3} \pm \frac{1}{3}\sqrt{22}$. When $x = -\infty$, $f(x)$ is $-$ and $f'(x)$ is $+$; hence the graph of $f(x)$ enters the finite field surrounding the origin from the *lower* left-hand quarter, whilst that of $f'(x)$ enters from the *upper* left-hand quarter.

Therefore when x is less than the least roots of both equations $f(x) = 0$ and $f'(x) = 0$, for instance, when $x = OA$ (Fig. 3), $f(x)$ and $f'(x)$ have unlike signs, as represented by the ordinates AB and AC. When

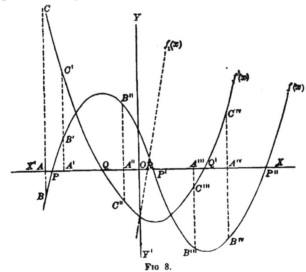

Fig. 3.

$x = -2 (= OP)$, $f(x)$ vanishes, *i.e.* $f(-2) = 0$, and the graph crosses the X-axis. $f(x)$ and $f'(x)$ now have like signs, as seen in the ordinates $A'B'$ and $A'C'$; and this is the case until x reaches a root (OQ) of

$f'(x) = 0$, when $f'(x)$ vanishes, and its graph crosses to the lower side of the X-axis. $f(x)$ and $f'(x)$ now have unlike signs until x reaches the next root ($OP' = 2 - \sqrt{2}$) of $f(x) = 0$, when $f(x)$ changes sign again. When x reaches the value $OQ'(= \frac{2}{3} + \frac{1}{3}\sqrt{22})$, $f'(x)$ again vanishes, and thus the signs are unlike until the last root of $f(x) = 0$ is reached, when the graph of $f(x)$ crosses the axis for the last time, and both graphs go off to $+\infty$ in the upper right-hand quarter.

Thus the graphs are seen to cross the axis alternately, the roots of the equation $f'(x) = 0$ being intermediate in value between those of $f(x) = 0$.

Also, if we construct the graph of the function $f_1(x)$, a straight line in this example, it is seen to cross the X-axis at $R(OR = \frac{6}{11})$, a point between the two points where the graph of $f'(x)$ crosses it.

The graphs of no two consecutive functions cross the axis at the same point; *i.e.* no two consecutive functions vanish for the same value of x, as shown in the demonstration.

It will be observed that, in the construction of these graphs, the unit of distance for vertical measurements is only about one-fourth the unit employed for horizontal distances. This is done in order to bring the figure within proper limits. The essential features of the curves, and hence their mutual relations, are preserved, however, and therefore the principles of the theorem are illustrated as well as if only one scale were used.

173. Resuming the series of functions,

$$f(x) = x^3 - 2x^2 - 6x + 4,$$
$$f'(x) = 3x^2 - 4x - 6,$$
$$f_1(x) = 11x - 6,$$
$$f_2(x) = +882,$$

we will employ Sturm's theorem in order to compare the results which it furnishes with those obtained above. The following table may be formed:

	$-\infty$	$+\infty$	0	-3	-2	-1	$+1$	$+2$	$+3$	$+4$
$f(x)$	$-$	$+$	$+$	$-$	0	$+$	$-$	$-$	$-$	$+$
$f'(x)$	$+$	$+$	$-$	$+$	$+$	$+$	$-$	$-$	$+$	$+$
$f_1(x)$	$-$	$+$	$-$	$-$	$-$	$-$	$+$	$+$	$+$	$+$
$f_2(x)$	$+$	$+$	$+$	$+$	$+$	$+$	$+$	$+$	$+$	$+$

The first column of signs presents the signs of the several functions when the functions are evaluated for $-\infty$; the second column, when they are evaluated for $+\infty$, and so on. In the $-\infty$ column, we note *three* variations, and in the $+\infty$ column, *no* variations; hence, since three variations have been lost, the equation $f(x)=0$ has three real roots. In the zero column, two variations appear; consequently one variation is lost between $-\infty$ and 0, and there is one negative root. In the column headed -3, there are three variations; therefore we know that the negative root does not lie between $-\infty$ and -3. Evaluating for -2, we observe that the number of variations is two; hence, since a variation has been lost between -3 and -2, we know that there is a root between -3 and -2; and since $f(-2)$ vanishes, this root is -2.

Since we have found one root exactly, we might use it to lower the degree of the equation, but in this case we will continue evaluating in order to exhibit more fully the use of Sturm's theorem. When $x=0$, two variations appear in the series of signs, but evaluating for $+1$ we obtain only one variation; therefore there is a root between 0 and $+1$.

186   ALGEBRA.

Continuing in this manner, it is found that the last variation is lost between $+3$ and $+4$; therefore the third root is situated between $+3$ and $+4$. We may continue this process, narrowing the limits between which any root is situated; thus we may evaluate for .5, and so determine whether the second root is between 0 and .5, or between .5 and 1. It is plainly possible to determine the roots to as many decimal places as may be desired, although the labor of evaluating increases as we go on. For Horner's Method of evaluating, see Todhunter's *Theory of Equations*.

174. A study of the above table of signs, and of other similar tables, shows that although no variation is lost except when x passes through a root of $f(x) = 0$, yet there is a shifting of the signs as x progresses through all values from $-\infty$ to $+\infty$; for, of course, the sign of any Sturmian function must change whenever x passes through a root (or an odd number of equal roots) of $f_i(x) = 0$, $f_i(x)$ representing any Sturmian function.

The demonstration of Sturm's theorem involves the assumption that the equation $f(x) = 0$ has no equal roots. It does not follow, however, that the theorem is inapplicable when we are not sure in advance that a given equation has no equal roots; for the process of forming the Sturmian functions is essentially the process of finding the H.C.F. of $f(x)$ and $f'(x)$; and if the given equation does contain equal roots, the last remainder, *i.e.* the last Sturmian function, will be zero, and the next to the last Sturmian function will be the H.C.F. In such a case, the equation $f(x) = 0$ may be reduced by dividing $f(x)$ by the product of the m linear factors $(x - \alpha)(x - \alpha) \cdots$, where α is one of the m equal roots;

STURM'S THEOREM. 187

or, if $f(x)=0$ has more than one set of equal roots, by dividing by the product of the different sets of linear factors. Suppose the quotient to be $\phi(x)$, so that we now have $\phi(x)=0$ for solution. Beginning *de novo*, we find $\phi'(x)$, and form the Sturmian functions. The series of functions $\phi(x)$, $\phi'(x)$, $\phi_1(x)$, $\phi_2(x)$, etc., evaluated in the usual manner, furnish the number and situation of the real roots of $\phi(x)=0$; these roots, together with the equal roots previously found, are the required real roots of $f(x)=0$. Thus the theorem is applicable in all cases.

175. It is evident that the functions in the series $f(x)$, $f'(x)$, $f_1(x)$, $\cdots f_{n-1}(x)$, will, in general, be $n+1$ in number; but in certain cases, owing to the absence of terms in the given function, some of the remainders will be wanting. This can occur only when the given equation has imaginary roots; for it is clear that, in order to ensure a loss of n variations in the series of functions during the passage of x from $-\infty$ to $+\infty$, all the functions must be present. Moreover, they must all take the same sign when $x=+\infty$, and alternating signs when $x=-\infty$. Since the leading term of an equation is taken with the positive sign, we may state the condition for the reality of all the roots of any equation as follows: *In order that all the roots of an equation of the nth degree may be real, the leading coefficients of all the Sturmian functions, n − 1 in number, must be positive.*

EXAMPLES.

1. Find by Sturm's theorem the number and situation of the real roots of the following equations:

    1. $x^3 + x - 3 = 0$.
    2. $x^3 - 7x + 7 = 0$.
    3. $x^3 - 15x - 5 = 0$.
    4. $x^4 - 3x^2 - 6x - 2 = 0$.
    5. $x^3 - 3x^2 + 2x - 6 = 0$.
    6. $2x^4 - x^3 - 6x^2 - x + 2 = 0$.
    7. $x^5 - 8x^4 + 26x^3 - 44x^2 + 40x - 16 = 0$.
    8. $x^4 - 3x^2 - 6x - 2 = 0$.

2. Extract the cube root of 30 by solving the equation
$$x^3 - 30 = 0.$$

3. Find the condition that the incomplete cubic
$$x^3 + ax + b = 0$$
shall have all of its roots real.

4. Show that the equation
$$x^3 - 3ax^2 + \frac{a^3}{2} = 0$$
has three real roots situated as follows:

    one root between $\frac{3}{8}a$ and $\frac{4}{8}a$,
    one root between $\frac{11}{4}a$ and $\frac{12}{4}a$,
    one root between $-\frac{3}{8}a$ and $-\frac{4}{8}a$.

Examine this equation with special reference to equal roots.

CHAPTER XXIII.

DETERMINANTS.

176. Suppose we have two linear equations in two unknown quantities, as

$$a_1x + b_1y + c_1 = 0,$$
$$a_2x + b_2y + c_2 = 0.$$

Multiplying the first by a_2 and the second by a_1, and subtracting the latter product from the former, we obtain

$$(a_2b_1 - a_1b_2)y + a_2c_1 - a_1c_2 = 0;$$

whence $\qquad y = \dfrac{a_2c_1 - a_1c_2}{a_1b_2 - a_2b_1}.\qquad$ See (9), Art. 77.

Similarly $\qquad x = \dfrac{b_1c_2 - b_2c_1}{a_1b_2 - a_2b_1}.$

Since the given equations are general, these values of x and y are formulas which may be used for finding the values of x and y in special linear equations in two unknown quantities. It will be observed that the two formulas have the same denominator, and that this denominator involves only the coefficients of x and y. Again, the numerator of the formula for x does not involve the coefficient of x, and the numerator of the formula for y does not involve the coefficient of y. Further, the six constants are involved in such a regular manner in these formulas that they may be sorted out and arranged in lines.

Thus the binomial $b_1c_2 - b_2c_1$ may be written $\begin{vmatrix} b_1 & b_2 \\ c_1 & c_2 \end{vmatrix}$;

and the binomial $a_1b_2 - a_2b_1$, $\begin{vmatrix} a_1 & a_2 \\ b_1 & b_2 \end{vmatrix}$;

so that
$$x = \frac{\begin{vmatrix} b_1 & b_2 \\ c_1 & c_2 \end{vmatrix}}{\begin{vmatrix} a_1 & a_2 \\ b_1 & b_2 \end{vmatrix}}.$$

When the function $a_1b_2 - a_2b_1$ is written in the above form, it is called a **determinant**. The quantities a_1, a_2, b_1, b_2, are called **constituents**; the lines of constituents, when read horizontally, are called **rows**; when read vertically, they are called **columns**. The determinant is said to be **developed** when it is written in its usual algebraic form. This is accomplished by multiplying together the constituents in the diagonal, beginning with the upper left-hand constituent, and subtracting from this the product of the constituents in the other diagonal. That is, by definition,

$$\begin{vmatrix} a_1 & a_2 \\ b_1 & b_2 \end{vmatrix} \equiv a_1b_2 - a_2b_1.$$

Developing in the same manner the determinant

$$\begin{vmatrix} a_1 & b_1 \\ a_2 & b_2 \end{vmatrix},$$

we have $a_1b_2 - a_2b_1$.

Hence *a determinant of the second order is unaltered in value when rows are changed into columns and columns into rows.*

Again, since

$$\begin{vmatrix} a_1 & a_2 \\ b_1 & b_2 \end{vmatrix} = a_1 b_2 - a_2 b_1 = -(a_2 b_1 - a_1 b_2) = -\begin{vmatrix} a_2 & a_1 \\ b_2 & b_1 \end{vmatrix},$$

it follows that *the interchange of the two columns changes the sign of the determinant;* and since

$$a_2 b_1 - a_1 b_2 = \begin{vmatrix} b_1 & b_2 \\ a_1 & a_2 \end{vmatrix},$$

the interchange of the two rows changes the sign of the determinant.

177. Passing now to the case of three linear equations as

$$a_1 x + b_1 y + c_1 z + d_1 = 0,$$
$$a_2 x + b_2 y + c_2 z + d_2 = 0,$$
$$a_3 x + b_3 y + c_3 z + d_3 = 0,$$

if y and z be eliminated by the usual algebraic processes, we have

$$(a_1 b_2 c_3 - a_1 b_3 c_2 + a_2 b_3 c_1 - a_2 b_1 c_3 + a_3 b_1 c_2 - a_3 b_2 c_1)x$$
$$+ d_1 b_2 c_3 - d_1 b_3 c_2 + d_2 b_3 c_1 - d_2 b_1 c_3 + d_3 b_1 c_2 - d_3 b_2 c_1 = 0;$$

hence

$$x = -\frac{d_1 b_2 c_3 - d_1 b_3 c_2 + d_2 b_3 c_1 - d_2 b_1 c_3 + d_3 b_1 c_2 - d_3 b_2 c_1}{a_1 b_2 c_3 - a_1 b_3 c_2 + a_2 b_3 c_1 - a_2 b_1 c_3 + a_3 b_1 c_2 - a_3 b_2 c_1}. \quad (1)$$

Analogous expressions may be obtained for y and z. Now the polynomials in these values of x, y, and z may be written in terms of determinants of the second order.

Thus considering the denominator

$$a_1 b_2 c_3 - a_1 b_3 c_2 + a_2 b_3 c_1 - a_2 b_1 c_3 + a_3 b_1 c_2 - a_3 b_2 c_1, \quad (2)$$

we may write it

$$a_1(b_2c_3 - b_3c_2) + a_2(b_3c_1 - b_1c_3) + a_3(b_1c_2 - b_2c_1), \quad (3)$$

or
$$a_1 \begin{vmatrix} b_2 & b_3 \\ c_2 & c_3 \end{vmatrix} - a_2 \begin{vmatrix} b_1 & b_3 \\ c_1 & c_3 \end{vmatrix} + a_3 \begin{vmatrix} b_1 & b_2 \\ c_1 & c_2 \end{vmatrix}. \quad (4)$$

This expression may be defined as equal to the determinant of the third order,

$$\begin{vmatrix} a_1 & a_2 & a_3 \\ b_1 & b_2 & b_3 \\ c_1 & c_2 & c_3 \end{vmatrix}. \quad (5)$$

The three determinants of the second order are called the **minors** of the constituents a_1, a_2, a_3. Each minor is seen to consist of constituents which do not occur in the row or column of the corresponding constituent. Thus the minor of a_3 is

$$\begin{vmatrix} b_1 & b_2 \\ c_1 & c_2 \end{vmatrix}.$$

If we let the three minors equal A_1, A_2, A_3 respectively, (4) may be written

$$a_1A_1 - a_2A_2 + a_3A_3. \quad (6)$$

A determinant of the third order may be developed by first writing it in terms of its minors as in (4), or it may be developed directly if we consider the composition of the original polynomial,

$$a_1b_2c_3 - a_1b_3c_2 + a_2b_3c_1 - a_2b_1c_3 + a_3b_1c_2 - a_3b_2c_1.$$

This polynomial is seen to consist of six terms, three of which are preceded by the plus sign, and three by the minus sign; the positive terms being $a_1b_2c_3$, $a_2b_3c_1$, $a_3b_1c_2$. Of these, the first is the product of the constituents in the diagonal of (5), beginning with the upper left-hand constituent; the second is the product of the constituents in a parallel diagonal beginning with the second constituent in the first column, this product being multiplied into the third constituent of the first row; the third consists of the third constituent of the first column into the product of the remaining two constituents in a diagonal parallel to the primary (or principal) diagonal. Similarly, we obtain the three negative terms, beginning with the diagonal containing the constituents c_1, b_2, a_3, and proceeding upward until each constituent in the first column is used, the minus sign being prefixed to each one of these three products.

Disregarding signs, each term is seen to consist of the *product of one constituent, and only one, from each column and each row.*

178. The polynomial

$$a_1b_2c_3 - a_1b_3c_2 + a_2b_3c_1 - a_2b_1c_3 + a_3b_1c_2 - a_3b_2c_1$$

may be written

$$a_1(b_2c_3 - b_3c_2) - b_1(a_2c_3 - a_3c_2) + c_1(a_2b_3 - a_3b_2),$$

or
$$a_1 \begin{vmatrix} b_2 & c_2 \\ b_3 & c_3 \end{vmatrix} - b_1 \begin{vmatrix} a_2 & c_2 \\ a_3 & c_3 \end{vmatrix} + c_1 \begin{vmatrix} a_2 & b_2 \\ a_3 & b_3 \end{vmatrix}, \quad (7)$$

which may be condensed into the third order determinant,

$$\begin{vmatrix} a_1 & b_1 & c_1 \\ a_2 & b_2 & c_2 \\ a_3 & b_3 & c_3 \end{vmatrix}. \quad (8)$$

Comparing this with (5), which is also equal to (2), it follows that *in a determinant of the third order rows may* be written for columns and columns for rows without changing the value of the determinant.

179. Expression (4) may be written

$$-a_1\begin{vmatrix}b_3 & b_2\\ c_3 & c_2\end{vmatrix} - a_2\begin{vmatrix}b_1 & b_3\\ c_1 & c_3\end{vmatrix} - a_3\begin{vmatrix}b_2 & b_1\\ c_2 & c_1\end{vmatrix},$$

which equals

$$-\begin{vmatrix}a_1 & a_3 & a_2\\ b_1 & b_3 & b_2\\ c_1 & c_3 & c_2\end{vmatrix}.$$

Hence

$$\begin{vmatrix}a_1 & a_2 & a_3\\ b_1 & b_2 & b_3\\ c_1 & c_2 & c_3\end{vmatrix} = -\begin{vmatrix}a_1 & a_3 & a_2\\ b_1 & b_3 & b_2\\ c_1 & c_3 & c_2\end{vmatrix}; \qquad (9)$$

that is, *if two adjacent columns, or rows, of the determinant are interchanged, the sign of the determinant is changed, but its value remains unaltered.*

If an *even* number of interchanges be made between adjacent columns, or rows, the sign of the determinant remains the same; but if an *odd* number of interchanges be made, the sign of the determinant is changed.

180. Suppose that two rows, or two columns, of a determinant are the same, as in

$$\begin{vmatrix}a_1 & b_1 & c_1\\ a_1 & b_1 & c_1\\ a_3 & b_3 & c_3\end{vmatrix}.$$

Let D be the value of this determinant. Interchanging the first and second rows, its value becomes $-D$; but the determinant is unaltered; hence $D = -D$, and therefore $D = 0$.

Hence, *if two rows, or two columns, of a determinant are identical, the determinant vanishes.*

181. *If each constituent in any row, or any column, of a determinant is zero, the determinant vanishes.*

For such interchanges may be made that the given zero row, or zero column, shall become the first row; then developing the determinant in terms of its minors, each minor has zero for its coefficient; hence the expression vanishes.

182. Let the coefficient of each minor in (4) be multiplied by n; then we have

$$na_1 \begin{vmatrix} b_2 & b_3 \\ c_2 & c_3 \end{vmatrix} - na_2 \begin{vmatrix} b_1 & b_3 \\ c_1 & c_3 \end{vmatrix} + na_3 \begin{vmatrix} b_1 & b_2 \\ c_1 & c_2 \end{vmatrix},$$

which equals n times the original polynomial.

Hence

$$\begin{vmatrix} na_1 & na_2 & na_3 \\ b_1 & b_2 & b_3 \\ c_1 & c_2 & c_3 \end{vmatrix} = n \begin{vmatrix} a_1 & a_2 & a_3 \\ b_1 & b_2 & b_3 \\ c_1 & c_2 & c_3 \end{vmatrix}; \qquad (10)$$

that is, *if all the constituents of any row, or any column, be multiplied by the same quantity, the determinant is multiplied by that quantity.*

183. If each constituent in any row or column consists of two terms, the determinant can be expressed as the sum of two determinants; for the determinant

$$\begin{vmatrix} a_1 + \alpha_1 & b_1 & c_1 \\ a_2 + \alpha_2 & b_2 & c_2 \\ a_3 + \alpha_3 & b_3 & c_3 \end{vmatrix}$$

can be written

$$(a_1 + \alpha_1)A_1 - (a_2 + \alpha_2)A_2 + (a_3 + \alpha_3)A_3,$$

or $\quad (a_1A_1 - a_2A_2 + a_3A_3) + (\alpha_1A_1 - \alpha_2A_2 + \alpha_3A_3),$

which equals

$$\begin{vmatrix} a_1 & a_2 & a_3 \\ b_1 & b_2 & b_3 \\ c_1 & c_2 & c_3 \end{vmatrix} + \begin{vmatrix} \alpha_1 & \alpha_2 & \alpha_3 \\ b_1 & b_2 & b_3 \\ c_1 & c_2 & c_3 \end{vmatrix}. \quad (11)$$

184. Since the value of (3) will not be changed if we add and subtract the same quantity, we may write in place of it

$$a_1(b_2c_3 - b_3c_2) + nb_1(b_2c_3 - b_3c_2)$$
$$+ a_2(b_3c_1 - b_1c_3) + nb_2(b_3c_1 - b_1c_3)$$
$$+ a_3(b_1c_2 - b_2c_1) + nb_3(b_1c_2 - b_2c_1);$$

but this expression equals

$$(a_1 + nb_1)A_1 - (a_2 + nb_2)A_2 + (a_3 + nb_3)A_3;$$

hence

$$\begin{vmatrix} a_1 & b_1 & c_1 \\ a_2 & b_2 & c_2 \\ a_3 & b_3 & c_3 \end{vmatrix} = \begin{vmatrix} a_1 + nb_1 & b_1 & c_1 \\ a_2 + nb_2 & b_2 & c_2 \\ a_3 + nb_3 & b_3 & c_3 \end{vmatrix}; \quad (12)$$

that is, *any multiple of any column, or row, may be added to any other column, or row, without changing the value of the determinant.*

185. If the equations in x, y, z, Art. 177, are supposed homogeneous, d_1, d_2, d_3 are each zero, and consequently the numerator of (1) is zero. Hence, if x is not zero, the denominator of (1) must also equal zero, and we have $x = \frac{0}{0}$, an indeterminate quantity.

Thus the condition that the three linear homogeneous equations

$$a_1x + b_1y + c_1z = 0,$$
$$a_2x + b_2y + c_2z = 0,$$
$$a_3x + b_3y + c_3z = 0$$

shall be consistent is

$$\begin{vmatrix} a_1 & b_1 & c_1 \\ a_2 & b_2 & c_2 \\ a_3 & b_3 & c_3 \end{vmatrix} = 0. \qquad (13)$$

186. In case we have *two* linear homogeneous equations in three unknown quantities, they may be treated as non-homogeneous equations in two unknown quantities. For the equations

$$a_1x + b_1y + c_1z = 0,$$
$$a_2x + b_2y + c_2z = 0,$$

may be written

$$a_1\frac{x}{z} + b_1\frac{y}{z} + c_1 = 0,$$

$$a_2\frac{x}{z} + b_2\frac{y}{z} + c_2 = 0,$$

in which the ratios $\dfrac{x}{z}$ and $\dfrac{y}{z}$ may be regarded as the unknown quantities.

Then, by Art. 176,

$$\frac{x}{z} = \frac{b_1c_2 - b_2c_1}{a_1b_2 - a_2b_1} \text{ and } \frac{y}{z} = \frac{a_2c_1 - a_1c_2}{a_1b_2 - a_2b_1};$$

hence $\qquad \dfrac{x}{b_1c_2 - b_2c_1} = \dfrac{y}{a_2c_1 - a_1c_2} = \dfrac{z}{a_1b_2 - a_2b_1}, \qquad (14)$

or
$$\frac{x}{\begin{vmatrix} b_1 & b_2 \\ c_1 & c_2 \end{vmatrix}} = \frac{y}{\begin{vmatrix} a_2 & a_1 \\ c_2 & c_1 \end{vmatrix}} = \frac{z}{\begin{vmatrix} a_1 & a_2 \\ b_1 & b_2 \end{vmatrix}};\qquad(15)$$

that is, the quantities x, y, and z are proportional to the determinants formed as shown in (15).

The preceding principles govern the ordinary operations performed on determinants of the second and third orders. From the manner in which the third order determinant was obtained, it is evident that there may be determinants of the fourth and higher orders. For a more extended treatment of determinants in general the student is referred to Burnside and Panton's *Theory of Equations*, Todhunter's *Theory of Equations*, and similar works.

EXAMPLES.

1. Find the value of the following determinants:

$$\begin{vmatrix} 1 & 2 & 3 \\ 2 & 3 & 4 \\ 3 & 4 & 5 \end{vmatrix};\quad \begin{vmatrix} -1 & -1 & 1 \\ -3 & 1 & -4 \\ 2 & -3 & -5 \end{vmatrix};\quad \begin{vmatrix} 4 & -1 & -2 \\ 0 & 3 & 0 \\ 3 & -7 & 4 \end{vmatrix}.$$

2. Find the value of x in the equation
$$\begin{vmatrix} 1 & 1 & 1 \\ a & x & c \\ b & b & x \end{vmatrix} = 0.$$

3. Show that
$$\begin{vmatrix} a_2 & a & 1 \\ b_2 & b & 1 \\ c_2 & c & 1 \end{vmatrix} = (a-c)(b-c)(a-b).$$

DETERMINANTS. 199

4. State the condition that the equations
$$Ax' + By' + C = 0,$$
$$Ax'' + By'' + C = 0,$$
$$Ax''' + By''' + C = 0,$$
shall be simultaneous and independent.

5. (*a*) Write formula (1), Art. 177, as the ratio of two determinants of the third order.

(*b*) By means of the result in (*a*), write the value of x in each of the following sets of equations:

$$x + y + z = 9, \qquad \tfrac{1}{2}x + \tfrac{1}{3}y + \tfrac{1}{4}z = 62,$$
$$x + 3y - 3z = 7, \qquad \tfrac{1}{3}x + \tfrac{1}{4}y + \tfrac{1}{5}z = 47,$$
$$x - 4y + 8z = 8. \qquad \tfrac{1}{4}x + \tfrac{1}{5}y + \tfrac{1}{6}z = 38.$$
$$x + \tfrac{1}{2}y = 100, \qquad ay + bx = c,$$
$$y + \tfrac{1}{3}z = 100, \qquad cx + az = b,$$
$$z + \tfrac{1}{4}x = 100. \qquad bz + cy = a.$$

6. Show that
$$\begin{vmatrix} b+c & a-b & a \\ c+a & b-c & b \\ a+b & c-a & c \end{vmatrix}$$
$$= \begin{vmatrix} b & a & a \\ c & b & b \\ a & c & c \end{vmatrix} - \begin{vmatrix} b & b & a \\ c & c & b \\ a & a & c \end{vmatrix} + \begin{vmatrix} c & a & a \\ a & b & b \\ b & c & c \end{vmatrix} - \begin{vmatrix} c & b & a \\ a & c & b \\ b & a & c \end{vmatrix}$$
$$= 3abc - a^3 - b^3 - c^3.$$

CHAPTER XXIV.

THEORY OF LOGARITHMS.

187. The **logarithm** of a quantity is the exponent with which a given fixed finite number called the **base** is to be affected in order to produce the quantity.

Thus if a be taken as the base of a system of logarithms, and we have $a^x = n$, then $x = \log_a n$, which is read: x equals the logarithm of n to the base a. Suppose the base is 4; then $2 = \log_4 16$, $\frac{1}{2} = \log_4 2$, etc.

188. Since logarithms are exponents, the laws of exponents hold in the theory of logarithms, and we have the following theorems:

1. *The logarithm of 1 is 0, whatever the base may be.*
If $a^x = 1$, $x = 0$; hence $\log_a 1 = 0$.

2. *The logarithm of the base itself is 1.*
If $a^x = a$, $x = 1$; hence $\log_a a = 1$.

3. *The logarithm of the product of two quantities is equal to the sum of the logarithms of the two quantities.*

Let p and q be the two quantities, and suppose $m = \log p$ and $n = \log q$, the base being a.

Then $\qquad a^m = p$ and $a^n = q$;

therefore $\qquad pq = a^m a^n = a^{m+n}$;

hence $\qquad \log_a(pq) = m + n$;

and since $m + n = \log_a p + \log_a q$, the theorem is established.

THEORY OF LOGARITHMS.

4. *The logarithm of the quotient of one quantity divided by another equals the logarithm of the dividend minus the logarithm of the divisor.*

As before, let $a^m = p$ and $a^n = q$;

then $\qquad \dfrac{p}{q} = \dfrac{a^m}{a^n} = a^{m-n}$;

hence $\qquad \log_a\left(\dfrac{p}{q}\right) = m - n = \log_a p - \log_a q.$

5. *The logarithm of any power of a quantity equals the logarithm of the quantity multiplied by the index of the power.*

Let $\qquad p = a^m$; then $p^r = (a^m)^r = a^{mr}$;

therefore $\quad \log_a(p^r) = mr = r\log_a p.$

6. *The logarithm of any root of a quantity equals the logarithm of the quantity divided by the index of the root.*

Let $\qquad p = a^m$; then $p^{\frac{1}{r}} = (a^m)^{\frac{1}{r}} = a^{\frac{m}{r}}$;

therefore $\quad \log\!\left(p^{\frac{1}{r}}\right) = \dfrac{m}{r} = \dfrac{1}{r}\log_a p.$

189. To find the relation of any two systems of logarithms, let a be the base of one system and let b be the base of the other. Let p be a quantity whose logarithm we take in each system. If n and n' are the logarithms of this quantity in the two systems, we have $a^n = p$ and $b^{n'} = p$; therefore $a^n = b^{n'}$.

Now let m be the logarithm of b to the base a; *i.e.* let $a^m = b$. Raising both members of this equation to the n'th power, $a^{mn'} = b^{n'}$; therefore $a^{mn'} = a^n$, and hence $mn' = n$,

or $\qquad\qquad \log_b p = \dfrac{\log_a p}{\log_a b}. \qquad\qquad (1)$

The translation of this equation affords a rule for obtaining the logarithms of numbers in a second system when the logarithms of those numbers are already known in one system.

$\dfrac{1}{\log_a b}$, the constant which connects the two systems, is called the **modulus** of the b-system.

If we have a third system whose base is c, then from (1) we may write

$$\log_c p = \frac{\log_a p}{\log_a c}; \qquad (2)$$

and eliminating $\log_a p$ from (1) and (2), we have

$$\frac{\log_b p}{\log_c p} = \frac{\dfrac{1}{\log_a b}}{\dfrac{1}{\log_a c}}.$$

Hence *the logarithms of the same number, taken in different systems, are proportional to the moduli of those systems.*

190. Logarithms to the base 10 are called **common logarithms**.* If we wish to find directly the logarithms of the numbers 2, 3, 4, etc., when 10 is taken as the base, we have to solve the series of equations:

$$10^x = 2, \quad 10^x = 3, \quad 10^x = 4, \text{ etc.}$$

The functions of x appearing in these equations belong to the class referred to in Art. 122 as transcendental: the

---

* Logarithms were invented by John Napier of Merchiston, Scotland. The work, entitled *Mirifici logarithmorum canonis descriptio*, in which Napier announced his discovery, was published in 1614. The system of common logarithms was introduced in 1617, by Henry Briggs, an English mathematician.

THEORY OF LOGARITHMS. 203

equations containing them are accordingly termed transcendental equations; as such, they only admit of approximate solution, although we can find the value of x to any desired degree of accuracy. However, instead of finding directly the common logarithms of 2, 3, 4, etc., it is usual first to calculate the logarithms in a system known as the *Napierian*, and then to calculate them for the common system by means of the principle of Art. 189.

191. We proceed now to the investigation of formulas which will enable us to construct a table of logarithms.

Let the function $\left(1+\dfrac{1}{n}\right)^{nx}$, in which n is taken >1, be developed by the binomial theorem, and we have

$$\left(1+\frac{1}{n}\right)^{nx}=1+nx\frac{1}{n}+\frac{nx(nx-1)}{\underline{|2}}\frac{1}{n^2}$$

$$+\frac{nx(nx-1)(nx-2)}{\underline{|3}}\frac{1}{n^3}+\cdots$$

$$=1+x+\frac{x\left(x-\dfrac{1}{n}\right)}{\underline{|2}}+\frac{x\left(x-\dfrac{1}{n}\right)\left(x-\dfrac{2}{n}\right)}{\underline{|3}}+\cdots. \quad (1)$$

Since this is true for all values of x, we may put $x=1$; thus obtaining

$$\left(1+\frac{1}{n}\right)^n=1+1+\frac{1-\dfrac{1}{n}}{\underline{|2}}+\frac{\left(1-\dfrac{1}{n}\right)\left(1-\dfrac{2}{n}\right)}{\underline{|3}}+\cdots. \quad (2)$$

But $\qquad\left(1+\dfrac{1}{n}\right)^{nx}=\left\{\left(1+\dfrac{1}{n}\right)^n\right\}^x;$

therefore series (1) is the xth power of series (2); that is,

$$1 + x + \frac{x\left(x-\frac{1}{n}\right)}{\lfloor 2} + \frac{x\left(x-\frac{1}{n}\right)\left(x-\frac{2}{n}\right)}{\lfloor 3} + \cdots$$

$$= \left[1 + 1 + \frac{1-\frac{1}{n}}{\lfloor 2} + \frac{\left(1-\frac{1}{n}\right)\left(1-\frac{2}{n}\right)}{\lfloor 3} + \cdots\right]^x. \quad (3)$$

If n be indefinitely increased, equation (3) becomes

$$1 + x + \frac{x^2}{\lfloor 2} + \frac{x^3}{\lfloor 3} + \cdots = \left(1 + 1 + \frac{1}{\lfloor 2} + \frac{1}{\lfloor 3} + \cdots\right)^x. \quad (4)$$

The series

$$1 + 1 + \frac{1}{\lfloor 2} + \frac{1}{\lfloor 3} + \frac{1}{\lfloor 4} + \cdots$$

is denoted by e; hence

$$e^x = 1 + x + \frac{x^2}{\lfloor 2} + \frac{x^3}{\lfloor 3} + \frac{x^4}{\lfloor 4} + \cdots; \quad (5)$$

writing cx for x, this becomes

$$e^{cx} = 1 + cx + \frac{c^2x^2}{\lfloor 2} + \frac{c^3x^3}{\lfloor 3} + \frac{c^4x^4}{\lfloor 4} + \cdots. \quad (6)$$

Since c is any constant, we may let $e^c = a$, so that $c = \log_e a$; substituting this value of c in (6), we have

$$a^x = 1 + x\log_e a + \frac{x^2(\log_e a)^2}{\lfloor 2} + \frac{x^3(\log_e a)^3}{\lfloor 3} + \cdots. \quad (7)$$

Equation (7) is known as the **exponential theorem.**

192. The series

$$1 + 1 + \frac{1}{\lfloor 2} + \frac{1}{\lfloor 3} + \frac{1}{\lfloor 4} + \cdots,$$

THEORY OF LOGARITHMS.

for which e stands, is of great importance, as it is the base of the primary (Napierian) system from which any other system is derived. Logarithms occurring in abstract mathematics are usually Napierian logarithms, whilst common logarithms are employed for the purposes of numerical computation.

193. An approximate value of e may be calculated in the following manner:

$$1 + 1 = 2.0000000$$
$$\frac{1}{\underline{2}} = .5000000$$
$$\frac{1}{\underline{3}} = .1666667$$
$$\frac{1}{\underline{4}} = .0416667$$
$$\frac{1}{\underline{5}} = .0083333$$
$$\frac{1}{\underline{6}} = .0013889$$
$$\frac{1}{\underline{7}} = .0001984$$
$$\frac{1}{\underline{8}} = .0000248$$
$$\frac{1}{\underline{9}} = .0000028$$
$$\frac{1}{\underline{10}} = .0000003$$
$$\therefore e = 2.7182819$$

To obtain the term $\dfrac{1}{\lfloor 3}$, we have only to divide the preceding term by 3; to obtain the term $\dfrac{1}{\lfloor 4}$, we divide the preceding term by 4; and so on. By including a greater number of decimal places in these quotients and a greater number of terms in the series, the value of e may be found to any desired number of decimal places.

194. Having the exponential theorem, we may now proceed to the expansion of $\log_e(1+x)$ in ascending powers of x. From (7), Art. 191,

$$a^y = 1 + y\log_e a + \frac{y^2(\log_e a)^2}{\lfloor 2} + \frac{y^3(\log_e a)^3}{\lfloor 3} + \cdots;$$

therefore,

$$\frac{a^y-1}{y} = \log_e a + \frac{y(\log_e a)^2}{\lfloor 2} + \frac{y^2(\log_e a)^3}{\lfloor 3} + \cdots$$

$$= \log_e a + y\left\{\frac{(\log_e a)^2}{\lfloor 2} + \frac{y(\log_e a)^3}{\lfloor 3} + \cdots\right\}. \quad (8)$$

If y be now diminished indefinitely, the terms containing y in the right-hand member of (8) will also diminish indefinitely; so that, as y approaches zero as its limit, we have

$$\frac{a^y-1}{y} = \log_e a. \quad (9)$$

Writing $1+x$ for a, (9) becomes

$$\log_e(1+x) = \frac{1}{y}\left\{(1+x)^y - 1\right\};$$

ng $(1+x)^y$,

$$\log_e(1+x) = \frac{1}{y}\left\{yx + \frac{y(y-1)}{\underline{|2}}x^2 + \frac{y(y-1)(y-2)}{\underline{|3}}x^3 + \cdots\right\}$$

$$= x + \frac{y-1}{\underline{|2}}x^2 + \frac{(y-1)(y-2)}{\underline{|3}}x^3 + \cdots. \quad (10)$$

But (10) is only another form for (9), and (9) supposes that y approaches zero as its limit; hence (10) reduces to the form,

$$\log_e(1+x) = x - \frac{x^2}{2} + \frac{x^3}{3} - \frac{x^4}{4} + \cdots. \quad (11)$$

This is the **logarithmic series.**

It will be observed that (11) is a diverging series when $x > 1$; hence it holds only when $x < 1$. However, we know, without applying the test for convergency, that (11) is true only when $x < 1$; for the derivation of this series involves the development of $(1+x)^y$, and we have seen in a former article, Chap. XVIII., that the development of $(1+x)^y$ holds only when $x < 1$.

195. Although the logarithmic series itself is divergent, we may obtain from it a series which is convergent, and which may be used in constructing a table of logarithms.

Let $-x$ be written for x in (11);

then $\quad \log_e(1-x) = -x - \dfrac{x^2}{2} - \dfrac{x^3}{3} - \cdots. \quad (12)$

Subtracting (12) from (11), we have

$$\log_e(1+x) - \log_e(1-x) = 2\left(x + \frac{x^3}{3} + \frac{x^5}{5} + \cdots\right);$$

that is, $\quad \log_e\left(\dfrac{1+x}{1-x}\right) = 2\left(x + \dfrac{x^3}{3} + \dfrac{x^5}{5} + \cdots\right). \quad (13)$

Let $x = \dfrac{1}{2z+1}$; then $\dfrac{1+x}{1-x} = \dfrac{z+1}{z}$.

Substituting these values in (13), we have,

$$\log_e\left(\frac{z+1}{z}\right) = 2\left(\frac{1}{2z+1} + \frac{1}{3(2z+1)^3} + \frac{1}{5(2z+1)^5} + \cdots\right);$$

and hence $\log_e(z+1)$

$$= \log_e z + 2\left(\frac{1}{2z+1} + \frac{1}{3(2z+1)^3} + \frac{1}{5(2z+1)^5} + \cdots\right). \quad (14)$$

196. Our object is to compute the Napierian logarithms of the numbers 2, 3, 4, etc. For such numbers, (14) is evidently a converging series; at the same time the relation between x and z is such that the requirement that x shall be less than unity is fulfilled as z is made successively equal to 1, 2, etc.

The computation of the Napierian logarithm of 2 will illustrate the use of this series. Making $z = 1$, we have

$$\log_e 2 = 2\left(\frac{1}{2+1} + \frac{1}{3(2+1)^3} + \frac{1}{5(2+1)^5} + \cdots\right).$$

Performing the indicated operations in a manner similar to that in the computation of e, we have

3	2.00000000		
9	.66666667	1	.66666667
9	.07407407	3	.02469136
9	.00823045	5	.00164609
9	.00091449	7	.00013064
9	.00010161	9	.00001129
9	.00001129	11	.00000103
9	.00000125	13	00000009
	.00000014	15	.00000001

$$\therefore \log_e 2 = .69314718.$$

THEORY OF LOGARITHMS. 209

Similarly, the logarithm of 3 may be computed by making $z = 2$.

197. It will be observed that it is only necessary to compute the logarithms of prime numbers, since the logarithm of a composite number equals the sum of the logarithms of its factors.

Thus $\log_e 10 = \log_e 5 + \log_e 2$. By making $z = 4$, and so computing $\log_e 5$, and then adding the result to $\log_e 2$ found above, we shall obtain $\log_e 10 = 2.30258508$.

198. In order to apply the theory of Art. 189 to the relation of Napierian logarithms and common ones, let a become the base of the Napierian system, and b the base of the common system.

Then $$\log_{10} p = \frac{\log_e p}{\log_e 10}.$$

Since $\log_e 10$ has been found to be 2.30258508, the modulus of the common system

$$= \frac{1}{\log_e 10} = \frac{1}{2.30258508} = .43429448$$

(approximately).

The Napierian logarithms, computed by the method of Art. 196, have only to be multiplied by $.43429448+$, and the table of common logarithms is formed.

199. It will be seen from the nature of logarithms that by their aid the operations of multiplication, division, involution, and evolution may be performed. To illustrate, let it be required to multiply 8 by 4 by means of logarithms to the base 2.

$$\log_2 4 = 2; \quad \log_2 8 = 3.$$
$$\log_2(8 \times 4) = \log_2 8 + \log_2 4 = 5.$$
$$\therefore 8 \times 4 = 2^5 = 32.$$

200. In the tables formed as described in the preceding articles there will be no logarithms of negative numbers; for it is evident that no value of x will render either of the functions e^x, 10^x negative.

However, the fact that negative numbers have no logarithms occasions no practical difficulty; for if such numbers enter into an operation, they may be treated as if they were positive, their signs being taken into account only in writing the final result.

Thus if it is required to multiply -8 by 4, we use the factor -8 as if it were positive, obtaining the numerical result, 32, as above; but the true product is, of course, -32, since one negative factor is involved.

201. Since

$$\left. \begin{array}{l} 10^2 = 100 \\ 10^1 = 10 \\ 10^0 = 1 \\ 10^{-1} = .1 \\ 10^{-2} = .01 \\ \text{etc.} \end{array} \right\} \text{we have to the base 10,} \left\{ \begin{array}{l} \log 100 = 2 \\ \log 10 = 1 \\ \log 1 = 0 \\ \log .1 = -1 \\ \log .01 = -2 \\ \text{etc.} \end{array} \right.$$

Hence the following propositions relating to common logarithms:

1. *The logarithm of any number between 1 and 10 is a fraction between 0 and 1.*

2. *The logarithm of any number having two integral places is $1 +$ some fraction; and in general the number*

THEORY OF LOGARITHMS. 211

composing the integral part of the logarithm is one unit less than the number of integral places in the quantity whose logarithm is given.

3. *The logarithm of a proper fraction is negative.*

4. *If one number is $\frac{1}{10}$ of another, its logarithm will be less by unity.*

For example, $\log 5 = .69897$ (carried to the fifth decimal place inclusive);

$\log \frac{5}{10} = \log 5 - \log 10 = .69897 - 1 = -.30103$.
$\log \frac{5}{100} = \log \frac{5}{10} - \log 10 = .69897 - 1 - 1 = -1.30103$.

202. Instead of combining .69897 and -1 in the above example, we may write $\bar{1}.69897$, which must be taken as equal to $-.30103$.

Similarly, we have

$$\log .05 = \bar{2}.69897,$$
$$\log .005 = \bar{3}.69897,$$
$$\log 50 = 1.69897,$$
etc.

To illustrate further, suppose we have a number consisting of several figures, as 3.274. Since this number is > 0 and < 10, its logarithm must be a proper fraction. From a seven-place table we find that $\log 3.274 = .5150787$. Employing the method used in the first example, we obtain

$$\log .03274 = \bar{2}.5150787,$$
$$\log\ .3274 = \bar{1}.5150787,$$
$$\log\ 3.274 = 0.5150787,$$
$$\log\ 32.74 = 1.5150787,$$
$$\log\ 327.4 = 2.5150787,$$
etc.

Generalizing, let l be the logarithm of any number as N where N is > 0 and < 10. Then if N be multiplied by any power of 10, as $(10)^r$, we have

$$\log (10)^r N = r + \log N = r + l;$$

and if N be multiplied by any power of .1, as $(.1)^r$,

$$\log (.1)^r N = - r + l.$$

In these formulas l is limited to positive fractional values, and r to positive integral values.

203. Logarithms are written as shown in the examples of the preceding article. The integral part is called the **characteristic**, and the decimal part the **mantissa**. The minus sign is placed above the characteristic rather than before it in order to indicate that the characteristic alone is negative. The propositions of Art. 201, together with the method just explained for writing negative logarithms, afford two important rules:

RULE 1. *The characteristic of the logarithm of an integral number, or of a mixed integral and decimal fractional number, is one less than the number of integral places in the number.*

RULE 2. *The characteristic of the logarithm of a number entirely decimal fractional is negative, and numerically one greater than the number of 0's immediately following the decimal point.*

204. It has been implied, Art. 199, that only four operations — multiplication, division, involution, evolution — can be performed by means of logarithms. It should be added, however, that tables based upon logarithmic tables have been made by means of which the operations of addition and subtraction may be performed. Zech's *Tafeln der Additions und Subtractions-Logarithmen* is of this description. Let it be required to add, or take the difference of, two numbers, a and b, where $a > b$. Let A represent the argument and F the function for addition, and A' represent the argument and F' the function for subtraction. To use Zech's *Tables* we have, then, the following formulas:

For addition $\begin{cases} \log a - \log b = A, \\ \log (a + b) = \log a + F. \end{cases}$

For subtraction $\begin{cases} \log a - \log b = A', \text{ or } F', \\ \log (a - b) = \log a - F', \text{ or } \log a - A'. \end{cases}$

Tables of addition and subtraction logarithms are of great service in certain astronomical computations in which the operations of addition and subtraction cannot be avoided.*

For explanation of the way in which logarithmic tables are to be used, the student is referred to the explanatory text which usually accompanies tables of logarithms.

EXAMPLES.

1. Given:
$$\log 2 = 0.3010300,$$
$$\log 3 = 0.4771213,$$
$$\log 7 = 0.8450980;$$

show that
$$\log .128 = \bar{1}.1072100,$$
$$\log 14.4 = 1.1583625,$$
$$\log 4\tfrac{2}{3} = 0.6690067,$$
$$\log \sqrt{\tfrac{35}{27}} = 0.0563521.$$

* See Oppolzer's *Lehrbuch zur Bahnbestimmung der Kometen und Planeten. Erster Band, zweiter Theil.*

214　*ALGEBRA.*

2. Given $\log 2$ and $\log 3$, find the value of x in the equation $3^{x-2} = 5$; also, find the value of x in the equation $5^x = 10^3$.

3. Show that $\log \dfrac{a}{b} = - \log \dfrac{b}{a}$, and translate the equation into a theorem.

4. Given $y = \dfrac{1}{\sqrt{\pi}} \epsilon^{-x^2}$; solve for x.

5. Given the expression $\dfrac{(\epsilon^a)^b}{(\epsilon)^{ab}}$; write it in its simplest form by means of logarithms.

In the article *Logarithms* in the *Encyclopædia Britannica*, the student will find a history of logarithms, and also much additional theory. The article *Tables* in the same Encyclopædia gives an account of all the important logarithmic tables which have been constructed, beginning with Briggs' *Arithmetica Logarithmica* (London, 1624).

CHAPTER XXV.

MATHEMATICAL REASONING.

205. Mathematics deals with propositions of the form:

If α is β, γ is δ. (1)

E.g. 'If two planes are perpendicular to a third plane, their intersection is normal to the third plane.'

'If $ax^2 + bx + c = 0$, $x = -\dfrac{b}{2a} \pm \sqrt{\dfrac{b^2}{4a^2} - \dfrac{c}{a}}$.'

Mathematical demonstration, whether short and simple or long and complex, consists in showing that the case of γ being δ is necessarily involved in the case of α being β.

The first part, if α is β, is marked by the various terms: **premise, condition, assumption, supposition, datum, hypothesis.***

This formal condition, if α is β, may be simple or compound; in the latter case it contains sub-conditions. Also,

---

* Modern inductive science employs the term 'hypothesis' in a sense so entirely different from that ordinarily assigned to it in mathematics, and the word is of so much more service as a term of science, that the interests both of mathematics and science would probably be advanced if it were no longer used in works on algebra and geometry. For an account of the *rôle* of the scientific hypothesis, the student is referred to Fowler's *Inductive Logic;* Gore's *Art of Scientific Discovery;* G. K. Gilbert's *Inculcation of Scientific Method* (American Journal of Science, vol. xxxi.).

it is frequently presented in the form of an adjective or a modifying phrase, as when we speak of an isosceles triangle or an equation of the nth degree.

The chief rules governing the condition are that it must not contradict any simultaneous condition, and it must accord with previously established conclusions of the branch of mathematics to which it belongs.

The part, γ is δ, is usually called the **conclusion**; it may express all or only a portion of that which is involved in the case of α being β.

In form, the conditional proposition of mathematics is identical with the conditional proposition of science:

$$\text{If } a \text{ is } b, c \text{ is } d; \qquad (2)$$

in which the part, if a is b, stands for **cause**, and the part, c is d, expresses **effect**. The parts: if α is β, if a is b, may be conveniently referred to as **antecedents**, while the parts: γ is δ, c is d, are termed **consequents**.

Aside from form, (1) and (2) have nothing in common. A 'datum' in mathematics can scarcely be said to be even analogous to a 'cause' in science. The tasks presented are also in great contrast: the student of mathematics seeking to understand necessary relations and conclusions, while the student of science is engaged in discovering the possible causes, and singling out the actual cause of a given effect, or in learning the effect of an actual or assumed cause.

206. The fallacy of affirming the consequent, c is d, and thence inferring the antecedent, a is b, is of frequent occurrence in the non-mathematical proposition (2). This fallacy usual arises from a failure to observe that the full form of (2) is:

$$\left.\begin{array}{c} \text{If } a_1 \text{ is } b_1 \\ \text{or} \\ \text{If } a_2 \text{ is } b_2 \\ \text{or} \\ \cdot\ \cdot\ \cdot\ \cdot \\ \text{or} \\ \text{If } a_n \text{ is } b_n \end{array}\right\} c \text{ is } d; \qquad (3)$$

and that we must have means for the elimination of all the antecedents except one; the one that remains being the cause, simple or compound, of c being d in the case in question. On the other hand, to affirm or assume the consequent, γ is δ, and thence infer the antecedent, α is β, in (1) does not necessarily involve a fallacy. This reversed form: if γ is δ, α is β, is evidently what is known in mathematics as the **converse**; as such it requires proof, although it may be observed that in many cases, if not in all, this proof is unnecessary provided that the conclusion, γ is δ, adequately expresses what is involved in α's being β; we may then begin with either as premise, and make the other the conclusion.

For example, suppose we have the proposition:

'If n lines are parallel, they make equal angles (θ) with a given line.'

This cannot be converted into the proposition:

'If n lines make equal angles (θ) with a given line, they are parallel.'

But the conclusion may be taken for condition, and the condition for conclusion, in the more complete statement:

'If n lines are parallel, they make equal angles (θ, θ', θ'') with three given lines.'

In reading the calculus the student will find examples of incomplete conclusions. Thus we have the theorem:

'If $f(x)$ is a maximum, $f'(x)$ equals zero.'

It does not follow that if $f'(x)$ equals zero, $f(x)$ is necessarily a maximum. But suppose the theorem stated as follows:

'If $f(x)$ is a maximum when $x = a$, the first derivative which does not vanish in the development of $f(a \pm h)$ is an even derivative, and is negative.'

The converse of this proposition is true, and requires no proof; or more properly speaking, it is proved as soon as the direct proposition is proved. See Williamson's *Differential Calculus*, Art. 138.

Forms (1) and (2) may be avoided by employing the equivalent form:

$$\text{The case of } \begin{cases} \alpha \text{ being } \beta \\ a \text{ being } b \end{cases} \text{ is the case of } \begin{cases} \gamma \text{ being } \delta \\ c \text{ being } d. \end{cases} \quad (4)$$

Finally, (4) may be expressed,

$$\text{All } P \text{ is } Q. \quad (5)$$

207. A proposition in form (5) is described as a **universal affirmative**. The subject is said to be **distributed**, whilst the predicate is, in general, not distributed. Thus,

All freshmen are undergraduates,

is a proposition referring to all freshmen, but not to all undergraduates. What we mean is, that all freshmen are some undergraduates. Evidently the subject cannot be taken for predicate, and predicate for subject, in another universal affirmative. We may only affirm that

Some undergraduates are freshmen;

thus converting a universal into a **particular** affirmative. The equivalence of forms (1), (4), (5) may be shown in an example as follows:

'If an equation is of the third degree, it has at least one real root.

'The case of an equation being of the third degree is the case of its having at least one real root.

'All equations of the third degree have at least one real root.'

Whenever all Q is referred to through all P, the proposition may be reversed, forming a new universal affirmative. If, then, we write a double form for (5), as

$$\text{All } P \text{ is } \begin{cases} \text{(some) } Q \\ \text{(all) } Q; \end{cases} \qquad (6)$$

we shall have the corresponding double form:

$$\left.\begin{array}{c}\text{Some } Q \\ \text{All } Q\end{array}\right\} \text{ is } P. \qquad (7)$$

Mathematics is chiefly concerned with propositions which conform to the second type under (6), *i.e.* propositions in which the predicate as well as the subject is distributed.

Thus in Art. 164, the theorem is established by showing that if $f(x)$ has equal roots, the H. C. F. of $f(x)$ and $f'(x)$ is itself a function of x, and the argument involves the proposition: 'All equations having equal roots are all equations such that the H. C. F. of the function and its first derivative is a function of x;' *i.e.* all P is all Q; and conversely, all Q is all P.

The student will find many universal affirmatives in the preceding pages, although they may not at first appear to be such. It is important to observe that a proposition of this description is under discussion whenever we employ the equation

$$f(x) = x^n + p_1 x^{n-1} + \cdots + p_{n-1} x + p_n = 0;$$

for, by using the arbitrary constants n, p_1, p_2, $\cdots p_n$, one equation is made to include all equations of all finite degrees in one unknown quantity; and in establishing any truth in regard to this equation, that truth is established for all equations of the class embraced in the general equation.

208. The process by which we discover the conclusions necessarily implied or involved in given conditions is described as **deductive** reasoning. The most common variety of such reasoning is characterized by the presence of two categorical propositions called **premises,** one of which must be universal, and one of which must be affirmative; from these a conclusion, either universal or particular, is drawn, or is said to follow.

As an example of this kind of reasoning, if we make the two affirmations:

All freshmen are undergraduates;

All freshmen study mathematics;

we are obliged to conclude that

Some undergraduates study mathematics.

Again, from the statements:

All logarithms are exponents;

.301030 is a logarithm;

it follows that

.301030 is an exponent.

As another illustration:

No determinants have an odd number of constituents;

$\begin{vmatrix} 2 & 0 \\ 1 & 3 \end{vmatrix}$ is a determinant;

therefore $\begin{vmatrix} 2 & 0 \\ 1 & 3 \end{vmatrix}$ has not an odd number of constituents.

MATHEMATICAL REASONING. 221

Here the first premise is a universal negative, the second premise a particular affirmative, whilst the conclusion is a particular negative. The student will easily add other combinations of premises to those given above. Thus a case might be given in which one premise should be a universal affirmative, and the other a particular negative.

209. Besides arguments composed entirely of simple categorical propositions, deductive reasoning includes certain other forms in which there appear categorical propositions united by a conjunction.

Thus if we affirm:

$$p \text{ is } q_1 \text{ or } q_2 \text{ or } \cdots \text{ or } q_n; \qquad (8)$$
$$p \text{ is } q_1;$$

we conclude that

$$p \text{ is not } q_2 \text{ or } q_3 \text{ or } \cdots \text{ or } q_n.$$

Again, from the statements:

$$p \text{ is } q_1 \text{ or } q_2 \text{ or } \cdots \text{ or } q_n; \qquad (9)$$
$$p \text{ is not } q_2 \text{ or } \cdots \text{ or } q_n;$$

it follows that

$$p \text{ is } q_1.$$

An illustration will occur to those who have read analytic geometry. In the study of the equation

$$ax^2 + 2hxy + by^2 + 2gx + 2fy + c = 0,$$

the function $h^2 - ab$ is formed. In any given equation of the second degree in x and y, $h^2 - ab$ is 0 or > 0 or < 0; suppose it is > 0; then it is not 0 or < 0.

The object in introducing in this connection any of these forms, whether categorical or not, is mainly to call attention to their constant occurrence in mathematics

For a detailed and systematic examination of deductive

reasoning, the student is referred to any of the standard text-books on the subject.

Besides the disguising of the universal affirmative, already referred to, the frequent suppression, or non-expression, of a premise should be observed. These suppressed premises are, however, none the less present and essential to a sound argument.

Finally, as regards the universal affirmatives of mathematics, the fundamental ones are the two axioms: 'Things equal to the same thing are equal to each other,' and 'The sums of equals are equal.' These are the κοιναί ἔννοιαι which stand at the head of Euclid's list. The other so-called axioms which relate to pure quantity may be derived from these elementary truths.

210. Occasionally a method other than the deductive is followed in order to establish a mathematical proposition. Thus in Art. 152, instead of taking an equation of the nth degree as in most of the other articles, an equation of the fourth degree has been used. Any truth established respecting the equation of that article will be general so far as the coefficients are concerned, but it will be particular as regards the degree of the equation. But from the operation performed on the fourth degree equation the student perceives what the result would be in the case of equations of the fifth, sixth, and higher degrees. He performs the operation potentially on equations of those degrees also, and reaches a general result, *i.e.* a truth concerning the equation of the nth degree, without actually using that equation in the process.

A similar generalizing occurs when, by examining the coefficients of the first few terms of the binomial theorem, we are able to express the law of their formation in the formula for the nth term. Reasoning of this kind has been called 'mathematical induction' because it bears a resemblance to some of the methods employed in inductive reasoning proper. The results of mathematical induction are general, but its material consists of particular truths.

In some instances, mathematics presents problems which suggest methods partaking of the nature of experiment. A simple case occurs in connection with the graph of $x + 2$, Art. 154. In that article it is stated that the graph is always a straight line when $f(x)$ is of the first degree, and the required graph was drawn accordingly; but independently of that statement, the student might have located a large number of graphic points, and then observed that the line connecting them was very nearly straight, becoming more so as the measurements were made more precise. The construction of the graphs of several special linear functions would doubtless have led to the conviction that *all* functions of the first degree have rectilinear graphs. But the universal proposition would not have been proven; the number of graphs constructed would have been insignificant compared with those left unconstructed, and no adequate reason could be assigned why the next graph should not be a curved line.

On the other hand, if the properties of the graph of the general linear function $ax + b$ could be discovered from a study of the function itself, and if it could be proven deductively that the graph of $ax + b$ is a straight line, there would be no need to raise the question of the nature of the graph of any particular linear function.

The peculiar power of analytic or co-ordinate geometry consists in the fact that being algebraic it deals with universals, and is enabled to employ the methods of algebra to establish universal affirmatives which admit of a geometrical interpretation.

In so far as the experimental determination of a graph is inductive in character, the above comparison of this method with the deductive may suggest that inductive reasoning is less conclusive or less important than deductive; but this single case is no just ground for an inference, even if the example in question fairly represented the nature of any true inductive process. To infer, on the above evidence, the inferiority of inductive methods would be to fall into the same fallacy that would be committed if we were to construct the graph of one special linear function, and thence infer, without other proof, that all linear functions have rectilinear graphs; the fallacy being the common one of hasty generalization from comparatively few instances.

For a discussion of the methods of inductive logic, and the reliability of the results of those methods, one should read Mill's *System of Logic*. The great importance of induction will be evident when it is realized that it furnishes deductive reasoning with premises. In this connection reference may again be made to the Euclidean axioms which have been spoken of as constantly occurring in mathematical argument. The question arises: How were they established? At present the best evidence supports the hypothesis that these axioms are themselves generalizations from the universal experience of mankind. If this theory is the true explanation of the origin of axioms, mathematics is no exception to the law that deductive reasoning presupposes inductively established truths and rests upon them.

EXAMPLES.

1. Supply premises for the conclusion:

'The equation $x^3 - 2x^2 + x + 1 = 0$ has at least one real root.'

2. 'If the coefficients of an equation are all real, imaginary roots enter it in conjugate pairs.' State this theorem in categorical form, and prove that both subject and predicate are distributed.

3. 'The opposite angles of any quadrilateral which can be inscribed in a circle are supplementary.' Examine this theorem with reference to necessity of proof of the converse.

4. Examine the theorem of Art. 175 with reference to necessity of proof of the converse.

5. Distinguish between the conditional forms in the following cases (a) and (b):

If two triangles are mutually equiangular,
If two triangles have their corresponding sides proportional,
If two triangles have their sides respectively perpendicular,
. . . .
} they are similar. (a)

If there is a strike among miners,
If mines become exhausted,
If new markets are opened,
. . . .
} the price of coal rises. (b)

6. Express the *reductio ad absurdum* method of proof symbolically, using a notation similar to that of (8) and (9), p. 221.

7. 'The general equation of the second degree in x and y represents a conic.' Is the predicate distributed or not?

8. Point out the error in the following:
A conic can be made to pass through any five points;
A parabola is a conic,
Therefore a parabola can be made to pass through any five points.

9. Supply premises for the conclusion:
'No ellipse has asymptotes.'

BIBLIOLIFE

Old Books Deserve a New Life
www.bibliolife.com

Did you know that you can get most of our titles in our trademark **EasyScript**™ print format? **EasyScript**™ provides readers with a larger than average typeface, for a reading experience that's easier on the eyes.

Did you know that we have an ever-growing collection of books in many languages?

Order online:
www.bibliolife.com/store

Or to exclusively browse our **EasyScript**™ collection:
www.bibliogrande.com

At BiblioLife, we aim to make knowledge more accessible by making thousands of titles available to you – quickly and affordably.

Contact us:
BiblioLife
PO Box 21206
Charleston, SC 29413

Printed in Great Britain by
Amazon.co.uk, Ltd.,
Marston Gate.